About Island Press

Island Press is the only nonprofit organization in the United States whose principal purpose is the publication of books on environmental issues and natural resource management. We provide solutions-oriented information to professionals, public officials, business and community leaders, and concerned citizens who are shaping responses to environmental problems.

In 2004, Island Press celebrates its twentieth anniversary as the leading provider of timely and practical books that take a multidisciplinary approach to critical environmental concerns. Our growing list of titles reflects our commitment to bringing the best of an expanding body of literature to the environmental community throughout North America and the world.

Support for Island Press is provided by The Nathan Cummings Foundation, Geraldine R. Dodge Foundation, Doris Duke Charitable Foundation, Educational Foundation of America, The Charles Engelhard Foundation, The Ford Foundation, The George Gund Foundation, The Vira I. Heinz Endowment, The William and Flora Hewlett Foundation, Henry Luce Foundation, The John D. and Catherine T. MacArthur Foundation, The Andrew W. Mellon Foundation, The Moriah Fund, The Curtis and Edith Munson Foundation, The New-Land Foundation, Oak Foundation, The Overbrook Foundation, The David and Lucile Packard Foundation, The Pew Charitable Trusts, The Rockefeller Foundation, The Winslow Foundation, and other generous donors.

The opinions expressed in this book are those of the author(s) and do not necessarily reflect the views of these foundations.

About SCOPE

The Scientific Committee on Problems of the Environment (SCOPE) was established by the International Council for Science (ICSU) in 1969. It brings together natural and social scientists to identify emerging or potential environmental issues and to address jointly the nature and solution of environmental problems on a global basis. Operating at an interface between the science and decision-making sectors, SCOPE's interdisciplinary and critical focus on available knowledge provides analytical and practical tools to promote further research and more sustainable management of the Earth's resources. SCOPE's members, forty national science academies and research councils and twenty-two international scientific unions, committees, and societies, guide and develop its scientific program.

SCOPE 64

Sustaining Biodiversity and Ecosystem Services in Soils and Sediments

Contents

Contents

Figures and Tables

Figures

Tables

Foreword

During the past couple of decades, two seemingly unrelated themes have emerged, sequentially, as new focal points in the study of ecological systems. In addition to adding to the fund of knowledge about ecosystem science, the results of these studies, in combination, have had a large impact on how knowledge about the status and operation of ecosystems is conveyed to nonspecialists. One of these themes is the concept that biodiversity, *per se,* plays a role in the functioning of ecosystems. The second theme is that of linking ecosystem processes to the delivery of services to society.

The prevailing studies of ecosystem functioning up to this period concentrated on understanding the interactions of the most fundamental ecosystem units, that is, trophic levels, and their operation in controlling the fluxes of carbon, energy, water, and nutrients. This simple but powerful paradigm enabled great advances in our understanding of ecosystem dynamics.

There has been a growing concern in recent times, however, with the increasing loss of species from ecosystems due to impacts of biotic disruptions of various kinds (hunting, selective harvesting, invasions), land-use change, and pollution. The question arose of how these losses of species were affecting ecosystem functioning. This issue resulted in a large-scale international effort to probe this question across the major biomes of the world. The initial effort relied mostly on information gathered for other purposes, but it nevertheless was useful for analyzing this new question. More recently there has been a shift toward experimental approaches to this question, with a considerable body of literature resulting.

At the end of the initial global synthesis, which was run under the auspices of the Scientific Committee on Problems of the Environment (SCOPE), it was concluded that the analysis was hampered by our lack of knowledge of underground biodiversity and processes. The decision to initiate a new study on belowground biodiversity and ecosystem functioning was the genesis of this book and much of the work that led to it. Further discussion, however, led to the innovation that is captured in this volume. Rather than merely looking at soil biodiversity, it was proposed that the analysis focus on below-surface processes, including terrestrial and aquatic. This was a very bold move, since it brought two very disparate scientific communities together, yet these communities, as was suspected early on, shared commonality in the processes that they studied. Bringing these communities together also had the important result of focusing on the linkages between below-surface terrestrial, freshwater, and marine systems.

As our knowledge of biodiversity and ecosystem functioning was accumulating, another somewhat independent line of inquiry was developing. This was the consideration of how the results of ecosystem processes produces the goods and services upon which human societies depend. This has proven to be a very powerful extension of our understanding of the role of biodiversity in ecosystem functioning in terms that are extremely relevant to policies relating to the use of biotic resources. This work has turned out to be intellectually challenging. It focuses attention on ecosystem linkages in a whole new manner. It forces us to look at the whole train of ecosystem processes that result in clean water, for example, including all of the biological, physical, and chemical processes involved. Importantly, in the larger analysis, it involves the assessment of how those processes that deliver clean water, in this example, can be compromised by the alteration of ecosystem properties resulting from management practices designed to deliver different services, which we also value, such as food.

There is another important thread that is captured in this book in addition to system functioning, services, and linkages: the vulnerability of these systems to continue to provide services under global changes. Thus this volume is innovative in its scope and important in its conclusions. An excellent team under the leadership of Diana Wall has produced a volume that no doubt will be a template for future syntheses as well as an important guidepost of the crucial research needs in this vital area of research.

Harold A. Mooney
Department of Biological Sciences
Stanford, California, USA

Preface

This book about the biodiversity in soil, freshwater sediments, and marine sediments and their role in the operation of ecosystems and provision of ecosystem services is a project developed under the Scientific Committee on Problems of the Environment (SCOPE). SCOPE's mandate is to assemble, review, and assess available data on human-made environmental change and the effects of these changes on people. In doing so, SCOPE has collected and synthesized scientific information on the complex and dynamic network of flows and interactions of the major biogeochemical cycles over the last three decades (SCOPE 61). When this project was started in 1996, one phase of the SCOPE Global Carbon Cycle project on the flow, interaction, and fate of carbon and other nutrients from land via lakes and rivers to deep oceans was being finalized (SCOPE 57), and a major synthesis of the transfer cycles and management of phosphorus in the global environment had been published, as had aspects of sulfur cycling in wetlands, terrestrial ecosystems, and associated water bodies (SCOPE 48, 51, 54). The first workshop on nitrogen cycling in the North Atlantic ocean and its watersheds was also underway (Howarth 1996). All of these projects, along with a workshop on biodiversity and ecosystem functioning (Schulze and Mooney 1994), drew attention to a lack of science-based knowledge in the area of soil and sediment biodiversity and their influence on ecosystem functioning, and the need for a thorough review and synthesis of the information available, as it posed a significant impediment to achieving sustainability.

Since then, nearly 100 scientists and students from more than 20 countries have voluntarily contributed their time and intellectual knowledge toward achieving a better understanding of soil and sediment biodiversity and biogeochemical cycling, and the consequence of loss on ecosystem services. The productivity of this group over the intervening eight years has been prodigious (see the SCOPE Soil and Sediment Biodiversity and Ecosystem Functioning [SSBEF] Committee Publications Resulting from Three Workshops in the back of this volume).

The SCOPE Committee on Soil and Sediment Biodiversity and Ecosystem Functioning held its final international workshop in Estes Park, Colorado, in the fall of 2002;

the results of their deliberations are presented in this volume as a synthesis of their current understanding.

Diana H. Wall
Natural Resource Ecology Laboratory, Colorado State University
Fort Collins, CO, USA

John W.B. Stewart
SCOPE Editor-in-Chief
Salt Spring Island, BC, Canada

SCOPE Secretariat
51 Boulevard de Montmorency, 75016 Paris, France
Véronique Plocq Fichelet, Executive Director

Literature Cited

For SCOPE volumes cited, please see the SCOPE series list on page 261.

Howarth, R.W., editor. 1996. Nitrogen Cycling in the North Atlantic Ocean and its Watersheds. Norwell, MA, Kluwer Academic Publishers.
Schulze, E.-D., and H.A. Mooney, editors. 1994. *Biodiversity and Ecosystem Function (Ecological Studies,* Vol. 99). Berlin, Springer-Verlag.

Acknowledgments

This book is a product of the final workshop of the SCOPE Committee on Soil and Sediment Biodiversity and Ecosystem Functioning (SSBEF) held in October 2002. Funding for the workshop and the publication of this book was graciously provided by a private, anonymous US foundation that also supported many of the other exciting interdisciplinary SSBEF workshops. For their support and dedication to advancing understanding on the role of a significant component of the world's biodiversity, I am deeply appreciative. The Ministries of Agriculture and the Environment, the Netherlands, also contributed funding for early SCOPE SSBEF workshops.

The SCOPE Soil and Sediment Biodiversity and Ecosystem Functioning Committee began in 1996. The Steering Committee, composed of Margaret Palmer, USA; T. Henry Blackburn, Denmark; Fred Grassle, USA; Patricia Hutchings, Australia; Timo Kairesalo, Finland; Isao Koike, Japan; Josef Rusek, Czech Republic; David Hawksworth, Spain; and especially Lijbert Brussaard, the Netherlands, was instrumental in the direction of the SSBEF synthesis workshops and publications. Holley Zadeh devoted countless hours to preparation of this book, and her energy, patience, excellent organizational and scientific editing skills strengthened this book, for which I am ever so grateful. Additional thanks go to Véronique Plocq Fichelet, Executive Director, SCOPE, and Susan Greenwood Etienne, SCOPE secretariat; the many reviewers who contributed their time to the quality of this book; to Lily Huddleson for her work on the cover design and many of the figures; and to Laurie Richards, Patti Orth, and Stella Salvo at the Natural Resource Ecology Laboratory. Gina Adams added greatly to the success of the SSBEF Committee over the years, by co-authoring proposals, organizing interactive workshops, and editing publications. Hal Mooney and Alan Covich gave me unlimited and frequent encouragement and added scientific focus. Wren Wirth energized and enabled a real contribution to knowledge on life below the surface. To all of these friends and colleagues, my sincere thanks.

Diana H. Wall

1

The Need for Understanding How Biodiversity and Ecosystem Functioning Affect Ecosystem Services in Soils and Sediments

Diana H. Wall, Richard D. Bardgett, Alan P. Covich, and Paul V.R. Snelgrove

There is an astonishing diversity of life in mud and dirt. Much of life as we know it is supported by the soils and sediments of freshwater lakes, ponds, streams, and rivers, and the vast sediments of estuaries and the ocean floor. Together, soils and sediments form an interconnected subsurface habitat that teems with millions of species providing essential ecosystem services for human well-being, such as cycling of nutrients, soil stabilization, and water purification. Like microbes, plants, and animals in more visible habitats across the earth, the below-surface habitats and their biodiversity are being modified at an unparalleled rate. Land use change and sediment change (trawling, dredging, damming, drying up of rivers), the movement and introduction of biotic species, changes in atmospheric composition (CO_2, increasing availability of fixed nitrogen), climate change, and pollution are all agents of change to subsurface ecosystems. These alterations of soil and sediment habitats, and their biota, have major consequences for humans. Soils, freshwater and marine sediments and their biota are non-renewable natural resources that humans depend on for the many goods and services that are so tightly linked to the economic basis of societies. Former US President Franklin D. Roosevelt once noted, "A nation that destroys its soils, destroys itself."

In the face of the rapid and massive transformation of global ecosystems, we need to assess the vulnerability of soil and sediment biodiversity to change and, in turn, to assess how this affects the nature of ecosystem services provided to humanity. This becomes increasingly important as we ask how we can sustainably conserve and manage these biota and habitats to ensure that future generations receive their critical ecosystem services.

This book addresses these and other important questions relating to how soil and sediment biodiversity contribute to human well-being and overall ecosystem function.

Freshwater Sediments, Marine Sediments, and Soils

Freshwater sediments, marine sediments, and soils cover the Earth's surface (Table 1.1) and are critical links between the terrestrial, aquatic, and atmospheric realms (Figure 1.1). These below-surface habitats are arguably the most diverse on the planet, teeming with a complex assemblage of species. The profusion of organisms and the composition of the biotic assemblages are integral to the maintenance of below-surface and above-surface habitats, to ecosystem functioning, and to the provision of ecosystem services that are crucial for human well-being (Wall et al. 2001b).

The public, farmers, gardeners, tourist industries, shoreline residents, and fishers are interested in the maintenance of sediments and soils for production of harvestable crops, recreation, and beauty of the landscape. These land users typically have high regard for the ecology of the habitats that they rely on for their livelihoods. They need to know: Will these shared resources—the soils and sediments and their biodiversity—be sustained in the future given increasing human populations and numerous and rapidly occurring changes in the environment? This has not been an easy question for scientists to answer.

Until recent decades, scientists considered the biota in the Earth's soils and sediments to be a "black box": They monitored the physical and chemical components of these environments, but treated the diverse, smaller organisms that comprise the soil and sediment community as an "unknown, undefined" set of functional groups. Now, facing this era of unprecedented anthropogenic disturbance resulting in biodiversity and habitat loss and the spread of invasive species, an urgent question facing both scientists and decision-makers is: Which taxa, and how much biodiversity, must be conserved to maintain or restore essential ecosystem functioning such as plant and animal production, breakdown of organic wastes and nutrient cycling?

Table 1.1. A general comparison of global characteristics of soils, freshwater sediments, and marine sediments.

Parameters	Soils	Freshwater sediments	Marine sediments
Global coverage	1.2×10^8 km^2	2.5×10^6 km^2	3.5×10^8 km^2
Carbon storage	1500 Gt	0.06 Gt	3800 Gt
Organic content	High	Low	Low
Oxygenation	Oxic	Oxic-anoxic	Oxic-anoxic
Salinity	Low	Low-High	High
Pressure	Low	Moderate	High

Modified from Wall Freckman et al. 1997.

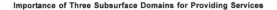

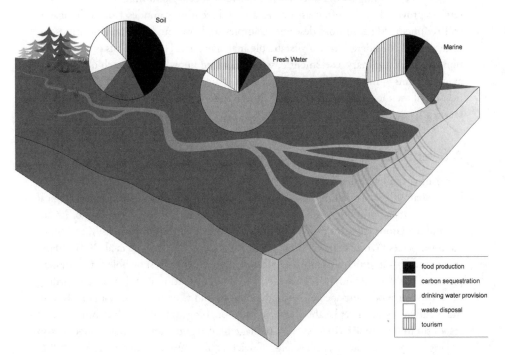

Figure 1.1. Schematic depiction of relative importance of different ecosystem goods and processes in terrestrial, freshwater, and marine sediments. The circles represent the relative contributions of the ecosystem services provided by soil, freshwater sediments, and marine sediments along the land-sea gradient.

The extremely diverse soil- and sediment-dwelling organisms occur all over the Earth and play critical roles in regulating the most vital ecosystem services (Daily et al. 1997; Petersen & Lubchenco 1997; Postel & Carpenter 1997). Identifying the significance of below-surface biological diversity for ecosystem functioning under scenarios of global change is being increasingly recognized as a major research priority (Lake et al. 2000; Smith et al. 2000; Wolters et al. 2000; Wall et al. 2001a). Changes in land-use practices affecting soils, in turn, have impacts for sediments of freshwater lakes, streams, rivers, estuaries, bays, and oceans. Alteration of hydrologic processes, contamination of soils, surface waters, and ground waters, and climate change are but a few examples of pressing problems that cannot be managed sustainably without a more complete understanding of the biota and ecosystem dynamics of soils and sediments. Applying this integrative knowledge to options for management and conservation will be crucial for long-term global sustainability of ecosystems and the welfare of human society (Millennium Ecosystem Assessment 2003).

This book highlights the fascinating biodiversity, ecosystem functions, and critical services provided by this often-assumed static and seemingly nondescript world of soils and sediments. The authors describe examples and present important priorities for research and considerations for sustainable management. This synthesis is an international, transdisciplinary assessment based on detailed knowledge for each of three separate domains—marine sediments, freshwater sediments, and soils—as well as identification of the most critical biota and their functions necessary for sustaining ecosystem services.

A Scientific Challenge

Furnishing the needed scientific information for this book has been challenging. We know that soils and sediments in most ecosystems contain enormous biological diversity—tens of millions of bacteria, thousands of fungal species, millions of protozoa and nematodes, up to a million arthropods (Groffman & Bohlen 1999), and thousands of species (World Conservation Monitoring Centre 1992; Wall et al. 2001a,b) per square meter—and that assemblages of these organisms are responsible for ecosystem processes (Tables 1.2 and 1.3). However, specific information is still lacking regarding the importance of many species-specific functions and whether the species involved in these functions are irreplaceable. Information on biogeographical distributions and responses to human activities, such as management regimes and global change, is also lacking for most keystone species (e.g., Covich et al. 1999; Levin et al. 2001; Wall et al. 2001a,b).

Below-surface organisms generally have been neglected by researchers, and often have been unrecognized by management and conservation programs relative to above-surface biota. Scientific research as a whole has understandably emphasized the visible: large plants and animals above the Earth's surface (both aquatic and terrestrial). For example, the latitudinal distribution of species, species ranges, and the occurrence of hot spots of biodiversity—all information used for local to national management and policy decisions—are based primarily on the larger, charismatic above-surface organisms. The emphasis on above-surface research continues in universities; there are fewer specialists trained to study the taxonomy, evolutionary biology, and biodiversity of below-surface organisms and their biogeochemical interactions, which limits our understanding in each domain compared with our understanding of above-surface domains. Separate scientific disciplines have developed based on "distinct habitats" in soils, freshwater systems, and marine systems, increasing understanding within each but effectively hindering the study of soils and sediments as an ecological continuum. For example, the biodiversity linkages between the soils of farms and cities in terms of runoff to freshwater sediments, and then to ocean sediments, have typically been neglected. Part of this neglect may be because of the additional complexity and cost involved in organizing interdisciplinary teams, or because of the manner in which research organizations have traditionally focused on specific disciplines. Scientific publications addressing the

Table 1.2. A comparison of the biodiversity (described species) in soil and marine sediments. Ants and earthworms are not present in marine sediments. Numbers of species of crustaceans in soils are for the Isopoda (Brusca 1997).

| Biota | Numbers of described species | |
	Soil	Marine sediments[b]
Fungi	18–35,000[a]	600
Protozoa	1,500[a]	2,000
Nematodes	5,000[a]	4,000
Ants	8,800[a]	——
Earthworms	3,600[a]	——
Crustaceans	5,000[c]	21,000

[a] Modified from Brussaard et al. 1997
[b] Modified from Snelgrove et al. 1997
[c] Modified from Wall et al. 2001a

Table 1.3. Examples of the diverse biota within functional groups are listed for a few ecosystem processes that are similar in soils and sediments. The functional groups play an important role in global transfers and biogeochemical cycling.

Organisms	Functional groups	Ecosystem processes
Vertebrates (gophers, lizards, beavers, whales); Invertebrates (oligochaetaes, polychaetaes, crustaceans, mollusks; echinoderms in sediments; ants, termites in soils), plant roots, macrophytes	Bioturbators, ecosystem engineers	Soil and sediment alteration and structure, laterally and to greater depths, redistribute organic matter and microbes
Plant roots, algae, diatoms	Primary producers	Create biomass, stabilize soils and sediments
Stoneflies in fresh water, decapods, millipedes	Shredders	Fragment, rip and tear organic matter, providing smaller pieces for decay by other organisms
Bacteria and fungi	Decomposers	Recycle nutrients, increase nutrient availability for primary production
Symbiotic (e.g., *Rhizobium*) and asymbiotic bacteria (e.g., *Cyanobacter*, *Azobacter*)	Nitrogen fixers	Biologically fix atmospheric N_2
Methanogenic bacteria, denitrifying bacteria	Trace-gas producers	Transfer of C, N_2, N_2O, CH_4 denitrification
Roots, soil organisms	CO_2 producers	Respiration, emission of CO_2

See Wall et al. 2001a for detailed listing of numbers of described and estimated species globally in soil taxonomic groups by size.

subject of the relationship of biodiversity and ecosystem functioning in either marine or freshwater sediments make up only about 1/100th of those published on above-surface terrestrial ecosystems. Consequently, taxonomists have described only a fraction of below-surface diversity (e.g., less than 0.1 percent of the marine species may be known; Snelgrove et al. 1997) and there is no site on Earth for which all the soil or sediment species present have been described. Thus a major component of the global ecosystem has been a minor part of analyses that consider how modifications of ecosystems resulting from global changes would affect ecosystem processes and human well-being.

Building the Foundation for This Book

The synthesis of available knowledge on the biodiversity and ecosystem functioning below surface became an international scientific priority in 1992 as a result of a workshop on biodiversity and ecosystem functioning (Schulze & Mooney 1994). In 1995, a SCOPE (Scientific Committee on Problems in the Environment) Committee on Soil and Sediment Biodiversity and Ecosystem Functioning (SSBEF) began to evaluate available data with the goal of providing policy makers with the scientific tools and information needed to promote sound management. How did this committee proceed to build the foundation for this book?

Through unprecedented collaboration between 70 international taxonomists and ecosystem scientists with expertise in soils, freshwater sediments, and marine sediments, the SSBEF committee developed state-of-the-art interdisciplinary syntheses and identified research and policy areas that need the most urgent attention. The committee held three extremely successful international workshops (see the SCOPE SSBEF committee publication list at the back of this volume, and summaries at the website http://www.nrel.colostate.edu/soil/SCOPE/scope.html). The workshops have resulted in 41 publications in journals read by scientists, managers, and policy makers. Additionally, the workshop syntheses have helped launch a new integrative discipline that crosses traditionally isolated disciplines (e.g., taxonomy, biogeochemistry, ecology), management, and domains (terrestrial, atmospheric, freshwater, and marine). This new scientific approach has contributed data to advance a more integrated and holistic understanding of Earth-system functioning and provides a foundation for this book. The SCOPE SSBEF syntheses identified major gaps in knowledge and research priorities in three overarching areas:

1. *The importance of soil and sediment biodiversity and ecosystem functioning within domains (soils, freshwater sediments, marine sediments).* Keystone functional groups (a group of species that has a much greater impact on an ecosystem process through impact on trophic relations than would be expected from its biomass) were identified within each domain. Comparisons among all domains revealed that, in general, the keystone ecosystem functions were strikingly similar in all domains (Table 1.3),

although the taxa involved were often remarkably different. The diversity of functions performed may be more important than the diversity of organisms for sustaining ecosystem processes. Thus, there appear to be universally important functions performed across all soil and sediment domains, and these contribute to vital ecosystem goods and services (Table 1.4). For example, in all domains many species are part of a complex food web that breaks down organic matter—by shredding, ripping, or dissolving it—and thus recycles soil nutrients to living plants and releases carbon dioxide and sequesters carbon (Table 1.3). Species in both soils and sediments filter particles from water and influence its flow, in turn cleansing and purifying water and playing a pivotal role in nutrient cycling and in the Earth's hydrological cycle.

2. *The importance of soil and sediment biodiversity and ecosystem functioning across domains.* Collaboration of scientists from all domains was crucial to elucidating the understanding that below-surface species and the processes they regulate do not operate in just one domain (e.g., exclusively in soils, freshwater sediments, or marine sediments). These species play a critical role by crossing and connecting domains, below and above surface, and thus regulating essential global cycles that contribute to the stabilization of Earth's climate and the maintenance of functioning ecosystems. Moreover, human disturbances in one domain can have cascading effects onto other domains. The SSBEF review committee found that integrated knowledge, research, and management of the domain interfaces was severely lacking. They set this area of interface exploration as a major research priority in order to maintain the high diversity and important functions in soils and sediments.

3. *Threats to soil and sediment biodiversity and their functioning.* The SSBEF workshops revealed that across all domains, global change poses a significant threat to below-surface biodiversity and the ecosystem functions they regulate. There was evidence across domains that land use change (including deforestation, overfishing, damming of rivers, agricultural intensification, pollution, and increased trampling), invasive species, and climate change can shift species composition, eliminate species, and reduce diversity at local to regional scales (Lake et al. 2000; Smith et al. 2000; Wolters et al. 2000; Wall et al. 2001b). The loss of populations and species at these local and regional scales threatens biodiversity, since it reduces genetic diversity, foreclosing opportunities for evolution and homogenizing biodiversity across the landscape. Whether species become globally extinct depends on the geographical extent of their range relative to that of the disturbance. But high-resolution data on the effects of a perturbation on soil and sediment biodiversity at large biogeographic scales is incomplete and was identified as another research priority. The SSBEF committee also determined that across all domains, current and predicted global change effects on below-surface biodiversity are, and will be, manifested largely through changes linked to above-surface habitats and biodiversity. These above-surface changes will transform the below-surface physical-chemical

environment, alter the transfer of nutrients and other resources belowground, and decouple species-specific interactions. These transformations may, in turn, have multiple, significant consequences for above-surface habitats and for their biodiversity and ecological processes.

Specific information was lacking on the vulnerability to human activities of the most important below-surface taxa and functions and their linkages above surface. How this vulnerability might be ameliorated by management options was considered an urgent priority for further research and synthesis. The workshops and syntheses of the SSBEF committee have advanced a paradigm shift in understanding biodiversity in soils and sediments. Scientists have acquired a more complete picture of these domains, especially the keystone taxa and functions, and are beginning to learn to what extent they are being, and may continue to be, disrupted and impaired by global change.

This improved, holistic understanding of below-surface biodiversity and of the linkages to organisms both above-surface and among domains was necessary before we could proceed toward addressing questions on the critical taxa, their ecosystem functions, habitats, biogeographical occurrence, and vulnerabilities.

How This Book Is Organized

Part I of this book outlines the critical soil and sediment ecosystem services (e.g., carbon sequestration, oxygen production, renewal of fertility, cleansing of water, provision of food; see Table 1.4) that sustain our natural and managed ecosystems. We also discuss, in Part I, the most essential below-surface habitats, ecological functions, and taxa for the provision of these services at different spatial and temporal scales. Table 1.4 provides a template of the types of ecosystem services that are expanded upon and discussed in each chapter. In Part II, ecologists provide a scientific appraisal of the vulnerability of the biota and functions in each domain and discuss how alterations resulting from global changes may affect the composition and operation of ecosystems and, in turn, the provision of ecosystem services. The integration of current knowledge revealed from the appraisal of each of the three domains—soils, freshwater sediments, and marine sediments—is then applied in Part III toward understanding the connectivity of the domains and the trade-offs that must be considered for managing and sustaining these systems.

This book provides an in-depth analysis of each domain and builds on previous reviews, examining ecosystem services (Daily et al. 2000; Dasgupta et al. 2000) and vulnerability of biodiversity to perform ecological functions in soils (Daily et al. 1997), and freshwater (Postel & Carpenter 1997) and marine ecosystems (Petersen & Lubchenco 1997). Wall et al. (2001b) considered soils and sediments as an ecological continuum and assessed the critical taxa and critical habitats for ecosystem services in terms of conserving these habitats. In this book we focus on the biodiversity within specific habitats

Table 1.4. Ecosystem services provided by soil and sediment biota.

Regulation of major biogeochemical cycles

Retention and delivery of nutrients to plants and algae

Generation and renewal of soil and sediment structure and soil fertility

Bioremediation of wastes and pollutants

Provision of clean drinking water

Modification of the hydrological cycle

 Mitigation of floods and droughts

 Erosion control

Translocation of nutrients, particles, and gases

Regulation of atmospheric trace gases (e.g., CO_2, NO_x)(production and consumption)

Modification of anthropogenically driven global change (e.g., carbon sequestration, modifiers of plant and algae responses)

Regulation of animal and plant (including algae, macrophytes) populations

Control of potential pests and pathogens

Contribution to plant production for food, fuel, and fiber

Contribution to landscape heterogeneity and stability

Vital component of habitats important for recreation and natural history

Modified from Daily et al. 1997; Wall and Virginia 2000; Wall et al. 2001a

of each domain, their vulnerablity, and how the vulnerability differs within habitats of each domain. We present case studies and trade-offs throughout to illustrate that the biodiversity and ecosystem functions with this ecological continuum below-surface must be considered for management of the global ecosystem.

Because there are frequently varying definitions for terminology, in this book we have used the following definitions, modified from the Secretariat of the Convention on Biological Diversity (2001), Chapin et al. 2002, and Folke et al. 2002, throughout the text:

- Biodiversity: the variability among living organisms including genetic, species, functional group, and ecosystem diversity across all temporal and spatial scales.
- Ecosystem Functioning: the activity of an ecosystem process, as in nitrogen mineralization rates in a given location and time period.
- Ecosystem Good/Product: substance directly produced by an ecosystem and used by people.
- Ecosystem Processes: inputs or losses of materials and energy to and from the ecosystem and the transfers of these substances among components of the system.
- Ecosystem Service: societally important consequences of ecosystem processes (e.g., water purification, mitigation of floods, pollination of crops).
- Function: ecosystem process.

• Functional Group/Type: a group of species that is similar with respect to their impacts on community or ecosystem processes; also defined with respect to their similarity of response to a given environmental change.
• Resilience: rate at which a system returns to its reference state after a perturbation.
• Resistance: the ability of an ecosystem or community property to withstand a major disturbance or stress.
• Vulnerability: The propensity of social and ecological systems to have diminished functional capacity following exposure to external stresses and shocks.

This book is the culmination of the SCOPE Committee on Soil and Sediment Biodiversity and Ecosystem Functioning. It has been an integrative activity that has resulted in a preliminary synthesis of knowledge on the below-surface world that will provide information needed for scientists, managers, and policy makers who are addressing options for sustainable management of ecosystems. However, all the authors involved in this final SSBEF synthesis realize that this is only an initial attempt to fill in the gaps on the contribution of the below-surface biodiversity and ecosystem functions to the provision of those ecosystem services necessary for human well-being. We are hopeful that the priorities and needs noted throughout the book will be an impetus for additional quantitative research and syntheses on the linkages of the wealth of biodiversity beneath and above the surface to the provision of ecosystem services.

Literature Cited

Brusca, R. 1997. Isopoda. Tree of Life Web Project, Tucson, Arizona. http://tolweb.org/tree?group=Isopoda&contgroup=Peracarida

Brussaard, L., V.M. Behan-Pelletier, D.E. Bignell, V.K. Brown, W.A.M. Didden, P.I. Folgarait, C. Fragoso, D.W. Freckman, V.V.S.R. Gupta, T. Hattori, D. Hawksworth, C. Klopatek, P. Lavelle, D. Malloch, J. Rusek, B. Söderström, J.M. Tiedje, and R.A. Virginia. 1997. Biodiversity and ecosystem functioning in soil. *Ambio* 26:563–570.

Chapin III, F.S., P.A. Matson, and H.A. Mooney. 2002. *Principles of Terrestrial Ecosystem Ecology.* New York, Springer-Verlag.

Covich, A.P., M.A. Palmer, and T.A. Crowl. 1999. The role of benthic invertebrate species in freshwater ecosystems. *BioScience* 49:119–127.

Daily, G.C., P.A. Matson, and P.M. Vitousek. 1997. Ecosystem services supplied by soil. In: *Nature's Services, Societal Dependence on Natural Ecosystems,* edited by G.C. Daily, pp. 113–132. Washington, DC, Island Press.

Daily, G.C., T. Soderqvist, S. Aniyar, K. Arrow, P. Dasgupta, P.R. Ehrlich, C. Folke, A. Jansson, B.-O. Jansson, N. Kautsky, S. Levin, J. Lubchenco, K.-G. Maler, D. Simpson, D. Starrett, D. Tilman, and B. Walker. 2000. The value of nature and the nature of value. *Science* 289:395–396.

Dasgupta, P., S. Levin, and J. Lubchenco. 2000. Economic pathways to ecological sustainability. *BioScience* 50:339–345.

Folke, C., S. Carpenter, T. Elmqvist, L. Gunderson, C.S. Holling, and B. Walker. 2002. Resilience and sustainable development: Building adaptive capacity in a world of transformations. *Ambio* 31:437–440.

Groffman, P.M., and P.J. Bohlen. 1999. Soil and sediment biodiversity: Cross-system comparisons and large-scale effects. *BioScience* 49:139–148.

Lake, P.S., M.A. Palmer, P. Biro, J. Cole, A.P. Covich, C. Dahm, J. Gibert, W. Goedkoop, K. Martens, and J. Verhoeven. 2000. Global change and the biodiversity of freshwater ecosystems: Impacts on linkages between above-sediment and below-sediment biota. *BioScience* 50:1099–1107.

Levin, L.A., D.F. Bosch, A. Covich, C. Dahm, C. Erseus, K.C. Ewel, R.T. Kneib, A. Moldenke, M.A. Palmer, P. Snelgrove, D. Strayer, and J.M. Weslawski. 2001. The function of marine critical transition zones and the importance of sediment biodiversity. *Ecosystems* 4:430–451.

Millennium Ecosystem Assessment. 2003. *Ecosystems and Human Well-Being: A Framework for Assessment.* Washington, DC, Island Press.

Peterson, C.H., and J. Lubchenco. 1997. Marine ecosystem services. In: *Nature's Services, Societal Dependence on Natural Ecosystems,* edited by G.C. Daily, pp. 177–194. Washington, DC, Island Press.

Postel, S., and S. Carpenter. 1997. Freshwater ecosystem services. In: *Nature's Services, Societal Dependence on Natural Ecosystems,* edited by G.C. Daily, pp. 195–214. Washington, DC, Island Press.

Schulze, E.-D., and H.A. Mooney, editors. 1994. *Biodiversity and Ecosystem Function (Ecological Studies,* Vol. 99). Berlin, Springer-Verlag.

Secretariat of the Convention on Biological Diversity. 2001. *Handbook on the Convention on Biological Diversity.* London and Sterling, VA, Earthscan Publications, Ltd.

Smith, C.R., M.C. Austen, G. Boucher, C. Heip, P.A. Hutchings, G.M. King, I. Koike, P.J.D. Lambshead, and P. Snelgrove. 2000. Global change and biodiversity linkages across the sediment-water interface. *BioScience* 50:1108–1120.

Snelgrove, P.V.R., T.H. Blackburn, P.A. Hutchings, D.M. Alongi, J.F. Grassle, H. Hummel, G. King, I. Koike, P.J.D. Lambshead, N.B. Ramsing, and V. Solis-Weiss. 1997. The importance of marine sediment biodiversity in ecosystem processes. *Ambio* 26:578–583.

Wall, D.H., and R.A. Virginia. 2000. The world beneath our feet: Soil biodiversity and ecosystem functioning. In: *Nature and Human Society: The Quest for a Sustainable World,* edited by P.R. Raven and T. Williams, pp. 225–241. Washington, DC, National Academy of Sciences and National Research Council.

Wall, D.H., G.A. Adams, and A.N. Parsons. 2001a. Soil biodiversity. In: *Global Biodiversity in a Changing Environment: Scenarios for the 21st Century,* edited by F.S. Chapin, III, O.E. Sala, and E. Huber-Sannwald, pp. 47–82. New York, Springer-Verlag.

Wall, D.H, P.V.R. Snelgrove, and A.P. Covich. 2001b. Conservation priorities for soil and sediment invertebrates. In: *Conservation Biology,* edited by M.E. Soulé and G.H. Orians, pp. 99–123. Washington, DC, Island Press.

Wall Freckman, D., T.H. Blackburn, L. Brussaard, P. Hutchings, M.A. Palmer, and P.V.R. Snelgrove. 1997. Linking biodiversity and ecosystem functioning of soils and sediments. *Ambio* 26:556–562.

Wolters, V., W.L. Silver, D.E. Bignell, D.C. Coleman, P. Lavelle, W.H. van der Putten,

P. de Ruiter, J. Rusek, D.H. Wall, D.A. Wardle, L. Brussaard, J.M. Dangerfield, V.K. Brown, K.E. Giller, D.U. Hooper, O. Sala, J. Tiedje, and J.A. van Veen. 2000. Effects of global changes on above- and belowground biodiversity in terrestrial ecosystems: Implications for ecosystem functioning. *BioScience* 50:1089–1098.

World Conservation Monitoring Centre. 1992. *Global Biodiversity: Status of the Earth's Living Resources*, edited by B. Groombridge. London, Chapman and Hall.

PART I

Ecosystem Processes and the Sustainable Delivery of Goods and Services

Alan P. Covich

In the next three chapters we describe how different types of goods and services are provided by diverse soil and sediment biota. We know that biotic communities living in soils as well as in freshwater and marine sediments provide many critical ecosystem services. Yet our understanding of the importance of biodiversity loss in these ecosystems is very limited. Information is available for some essential keystone species, but the many species-specific relationships that characterize these communities and their particular functional roles are just beginning to be studied. We discuss how soil and sediment biodiversity can influence rates of productivity, nutrient transformations, and decomposition of organic matter. We include case studies that demonstrate the degree to which the loss of a single species could alter these ecosystem services.

Soils are associated with critical processes, such as nutrient storage and cycling, that control production of agricultural crops, as well as natural plants that provide foods for people and a wide range of herbivores. Marine sediments are rich in species that also are essential for nutrient cycling and that provide foods used by many people and diverse food webs. Fresh waters provide unique ecosystem services, such as natural purification of drinking water, through a series of complex interactions among microbes and invertebrates that break down organic matter and pollutants such as excessive nutrients. Fish and shellfish production in oceans, lakes, and rivers clearly depend on the growth of various life stages of fishes that feed on a wide range of sizes and types of benthic prey species in complex food webs.

The sustainability of ecosystem services is linked to natural and human-derived sources of nutrients from surrounding agricultural fields, urban areas, and natural watersheds. The capacity of soils and sediments to store organic carbon and to buffer chemical transformations is a service that connects to many terrestrial and aquatic ecosystem dynamics. If nutrient inputs enter soils and sediments in appropriate proportions and concentrations from natural and human-derived sources, food webs function predictably and effectively to maintain ecosystem services. However, if managers focus only on crop or fish production, they may lose important values derived from related ecosystem functions that are critical for disease control and recreational uses of downstream waters. We stress that the overexploitation of one ecosystem service can lead to a disservice, a loss, or a reduction in benefit from another ecosystem service.

Providing a framework to disentangle and analyze the relative importance of interrelated ecosystem services is still very much needed. Such a framework could help managers avoid mistakes of linear analysis and focus on single-service attributes of biologically complex ecosystems. This process of evaluation requires interdisciplinary studies by ecologists and economists to determine relative as well as absolute values of ecosystem services. Trade-offs in managing for various combinations of ecosystem services are inherently complex.

In the next three chapters we discuss how sustainable goods and services are related to natural ecosystem processes. Ecologists, economists, and engineers are beginning to work together to determine how natural bioturbation and biofiltration in ecosystems work in conjunction with improved technological processes to break down organic matter or to manage agricultural production. The loss of diverse microbes and benthic invertebrates can result in a buildup of high concentrations of organic matter or can lock up essential nutrients that are needed to maintain production. In some cases, the sequestration of organic matter is a service if it reduces carbon dioxide in the atmosphere and reduces threats of global warming. In other cases, recycling of carbon and associated nutrients is an essential service. Natural processes associated with the "self-cleaning" of freshwater and marine ecosystems thus can maintain clean water, improved fishing, and sustainable delivery of ecosystem services. Moreover, these natural ecosystem processes can save money by eliminating the need for larger investments in engineered solutions for restoring or duplicating natural processes. However, it is also clear that natural ecosystem services have their limits when very large amounts of waste are allowed to enter and overload the ecosystem, or when toxic elements are dumped into natural ecosystems resulting in collapse of the entire system. In this section, each chapter identifies these complexities and summarizes the priorities for future research, and also suggests where additional understanding is needed to enhance policy.

2

The Sustainable Delivery of Goods and Services Provided by Soil Biota

Wim H. van der Putten, Jonathan M. Anderson, Richard D. Bardgett, Valerie Behan-Pelletier, David E. Bignell, George G. Brown, Valerie K. Brown, Lijbert Brussaard, H. William Hunt, Phillip Ineson, T. Hefin Jones, Patrick Lavelle, Eldor A. Paul, Mark St. John, David A. Wardle, Todd Wojtowicz, and Diana H. Wall

Soil systems provide numerous goods and services that are critical for human society (Daily et al. 1997; Wall et al. 2001; Millennium Ecosystem Assessment 2003). Many of these goods (e.g., food production, construction materials) and services (e.g., renewal of fertility, carbon sequestration, storage and purification of water, suppression of disease outbreaks) are in part mediated by soil organisms. The reduction of soil services as a consequence of improper management is rarely considered when evaluating consequences of management strategies and decisions. The consequence of such impropriety has been estimated to be in excess of US$1 trillion per year world-wide (Pimentel et al. 1997; Table 2.1), but the local financial consequences will vary according to how dependent the local economy is on ecosystem services. The role of soil organisms depends not only on the type of organisms present and their activity, but probably also on their diversity (Wardle 2002) and on a range of abiotic factors, some of which act locally (soil fertility), while others are more global (climate).

The management of soil systems requires an understanding of the underlying ecosystem processes and how these are influenced by the environment. This sets the borders for the ecology of the species involved. In this chapter, we provide a framework for assessing the role of soil organisms in the delivery of ecosystem goods and services. We use examples from grasslands, forests, and agriculture to illustrate how the consequences of management may be evaluated for the soil system in different environ-

Table 2.1. Total estimated economic benefits of biodiversity with special attention to the services that soil organism activities provide worldwide (modified from Pimentel et al. 1997).

Activity	Soil biodiversity involved in the activity	World economic benefits of biodiversity ($\times$ US10^9/year)
Waste recycling	Various saprophytic and litter feeding invertebrates (detritivores), fungi, bacteria, actinobacteria, and other microorganisms	760
Soil formation	Diverse soil organisms, e.g., earthworms, termites, fungi, eaubacteria, etc.	25
Nitrogen transformations	Biological nitrogen fixation by diazotroph bacteria, conversion of NH_4 to NO_3 by nitrifying bacteria, conversion of NO_3 to N_2 by denitrifying bacteria	90
Bioremediation of chemicals	Maintaining biodiversity in soils and water is imperative to the continued and improved effectiveness of bioremediation and biotreatment	121
Biotechnology	Nearly half of the current economic benefit of biotechnology related to agriculture involves nitrogen-fixing bacteria, pharmaceutical industry, etc.	6
Biocontrol of pests	Soil provides microhabitats for natural enemies of pests, soil organisms (e.g., mycorrhizae) that contribute to host plant resistance and plant pathogen control	160
Pollination	Many pollinators may have edaphic phase in their life cycle	200
Wild food	For example mushrooms, earthworms, small arthropods, etc.	180
Total		1,542

ments. We compare unmanaged with managed grassland and forest ecosystems, and non-tilled with tilled agricultural systems. We discuss which soil systems, habitats, ecological soil functions, soil taxa, and underlying processes are critical to the sustainability of delivering goods and services, and how human activities may affect these by land management and by inducing global change. We work along a gradient, where plant-soil feedbacks in unmanaged systems may resemble those in low-input agriculture or forestry. We discuss whether these continua do indeed exist and what may be learned from unmanaged ecosystems when attempting to manage or enhance the sustainability of managed systems.

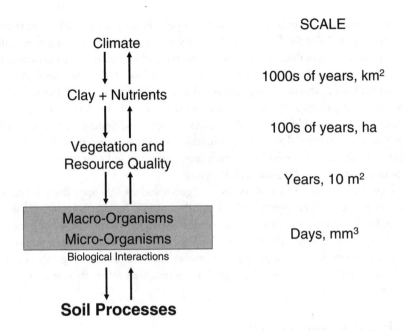

Figure 2.1. Hierarchy of the determinants of soil processes that provide ecosystem services (after Lavelle et al. 1993). Note that these determinants occur at different temporal and spatial scales.

Hierarchy of Environmental Controls of Soil Ecosystem Goods and Services

The contribution of soil organisms to ecosystem goods and services is determined by a suite of hierarchically organized abiotic factors, and by the nature of the plant community (Lavelle et al. 1993; Figure 2.1). At the highest level of the hierarchy, climate determines soil processes at regional and global scales (Gonzalez & Seastedt 2001). Climatic limitations, such as drought or low temperatures, directly determine the rates of the main physical, chemical, and microbiological reactions (Lavelle et al. 1997). At the second level in the hierarchy, the landscape and the original nature of parent material largely influence soil ecosystem goods and services. Nutrient heterogeneity of the substrate, along with the amount and quality of clay minerals, are the most important characteristics.

The third level in the hierarchy is the quality and quantity of the organic matter produced, and this depends on the nature and composition of plant communities. The release of energy and nutrients stored in dead organic matter depends strongly on the proportion of support tissues rich in lignin and lignocellulose; the proportion of secondary chemical compounds, such as tannins or polyphenols; and the ratio of nutrients to carbon (Grime 1979; Swift et al. 1979). Production of secondary products potentially affecting the decomposability of the organic matter may be enhanced by nutrient deficiencies and/or attack by herbivores (Waterman 1983; Baas 1989). The chemical composition of plants is, therefore, an important determinant of the composition and activity of the soil organism community. At the lowest hierarchical level, the soil organisms themselves operate within functional domains. A functional domain comprises a subset of the soil community that has similar functions or effects on, for example, soil structure (Lavelle 2002). The influence of each factor affecting soil ecosystems varies over temporal scales, from millimeters to kilometers.

Information on the sustainable delivery of goods and services originating from one region may not be directly applicable to another region; information on intensively managed soils, where functional domains differ considerably from those under extensive management, should not be generalized *a priori*. Therefore, our comparison of ecosystem management strategies for sustainable delivery of ecosystem goods and services as provided by soils should be regarded as a template, rather than as results that apply to the whole globe.

Biotic and Abiotic Drivers of the Underlying Processes in Soils

The complete destruction of the community of soil organisms—for example, due to erosion—results in obvious loss of soil ecosystem functions. Far less is known about the consequences of the loss of soil biodiversity for the sustainable delivery of ecosystem goods and services (Wall et al. 2001). Results from empirical studies on the relation between soil biodiversity and ecosystem functions range from positive to neutral or even negative (Mikola et al. 2002). However, it is obvious that soil biodiversity may act as a source of insurance, thereby making systems more stable. For example, when soil biodiversity is reduced by one stress factor or event, the soil community may be less able to recover from a repeated or second stress. Such accumulation of stresses may well result in the loss of stability of soil ecosystem processes (Griffiths et al. 2000).

Belowground processes that involve decomposition of organic matter, transformations of nutrients, and the supply of nutrients from the soil for plant growth are driven in the first instance by the activities of bacteria and fungi (Wardle 2002). Biotic drivers of microbially driven processes in soils include plants, their herbivores and pathogens, and other soil animals. Plant species differ in the quantity and quality of litter (dead organic matter that is sufficiently intact to be recognized) and rhizosphere materials that they return to the soil. This in turn governs the composition, growth, and activity of the microflora, and, hence, the rates of soil process (Hooper & Vitousek 1998). Aboveground herbivores can strongly influence soil organisms through a number of mechanisms by which they alter the quantity and quality of resources entering the soil (Bardgett & Wardle 2003). Soil animals have important effects on microbial activity at a range of scales (Lavelle 1997): by consuming microbes directly, by transforming litter and thus altering its physical structure, by creating biogenic structures, or by entering into mutualisms with microbes either in the "external rumen" or in the gut cavity (Lavelle 1997).

Some groups of soil organisms have a direct relationship with plant roots (Wardle et al. 2004). Soil pathogens and root herbivores are notorious for causing yield reductions of arable crops, re-sowing failures in grassland, disease in orchards that requires replanting, and die-back in production forest. Root pathogens and herbivores can also have strong impacts on the species composition of natural vegetation and the rate of changes therein (Brown & Gange 1989; van der Putten 2003). Root pathogens and herbivores contribute to primary and secondary plant succession (Brown & Gange 1992; van der Putten et al. 1993; De Deyn et al. 2003) and plant species diversity (Bever 1994; Packer & Clay 2000). Plant species that escape from soil pathogens may become invasive in new territories (Klironomos 2002; Reinhart et al. 2003). Root symbionts, such as mycorrhizal fungi and nitrogen fixing microorganisms (e.g., *Azobacter*, *Rhizobium*, and *Cyanobacter*), are important for plant nutrition in unmanaged systems and in non-tilled, low-input arable systems (Smith & Read 1997).

The various biotic and abiotic drivers in soil systems do not operate independently of one another, and interactions among them are important determinants of soil processes. For example, a recent litter exclusion study found that soil microarthropods play an important role in decomposition in a tropical forest but not in a temperate forest, where the abiotic drivers of microbial activity emerged as being of greater importance (Gonzalez & Seastedt 2001). Similarly, substrate fertility determines the role of microbially driven processes in supplying nutrients: in fertile conditions, plants produce litter of high quality from which ammonium (NH_4^+) is readily released by microbes and taken up by plants, while in infertile conditions plants produce poor-quality litter protected by polyphenols from which nitrogen is not readily mineralized. In these infertile systems, plants often bypass the mineralization process entirely by taking up organic nitrogen directly (e.g., Northup et al. 1995).

Context Dependence of Environmental Controls

Environmental controls of soil ecosystem processes and the delivery of soil ecosystem goods and services depend on climate, soil type, vegetation, and the local context (Figure 2.1). Stressed systems, whether they are stressed from extreme climatic seasonality (seasonal forests and savanna) or human interventions (managed grassland, managed forest, and tilled arable land), are usually characterized by lower diversity of soil fauna (there is little comparative information on microbes) and greater dominance of certain species (see Tables 2.2–2.4). Under these conditions the effects of key organisms, sometimes at species level, become more apparent (Anderson 1994). In arable soils, soil tillage consistently reduces the abundance of animals with large body sizes, whereas microbes and microfauna are not affected so much (Wardle et al. 1995). Here, we illustrate the consequences of the local context by discussing ecosystem processes, goods and services for three major types of land use: grassland, forest, and arable land and the role of soil organisms in controlling ecosystem goods and services (see Tables 2.2–2.4 for an overview of processes, goods, and services for grassland, forest, and arable land, respectively). Later, we discuss management trade-offs for each of these three systems and provide a conclusion on the importance of soil biodiversity.

Grasslands

Grasslands, including steppes, savannas, prairies, and tundra, are important terrestrial ecosystems covering about a quarter of the Earth's land surface. Grasslands build soil systems that differ from those of forests and other vegetation types. A key feature of grasslands, especially those that are more fertile, is their high turnover of shoot and root biomass, and the consequent large pool of labile organic matter at the soil surface. In contrast to many terrestrial ecosystems, heavy herbivore loads, both above- and belowground, are also a characteristic feature of fertile grasslands. This significantly influences plant growth and species composition, and the structure of grassland, since herbivores consume high proportions of annual aboveground (McNaughton 1983) and belowground (Stanton 1988) net primary production.

Aboveground herbivores also have a profound influence on nutrient pathways in grasslands through short circuiting both litter return and soil recycling processes (Bardgett & Wardle 2003). A large percentage of nutrients taken up by plants in grazing ecosystems is cycled directly through animal excreta, resulting in accelerated soil incorporation, particularly of nitrogen and phosphorus (Ruess & McNaughton 1987). Therefore, soils of grazed grasslands tend to have relatively small amounts of litter on the soil surface, but large amounts of organic nitrogen and carbon in the soil. These features combine to produce a soil environment that sustains, or is sustained by, an abundant and diverse faunal and microbial community.

In high latitude and high altitude grasslands, as well as in grasslands of semi-arid regions, soil processes and ecosystem goods and services are likely to be controlled more by abiotic factors than by biotic ones. This may be different in lowland temperate or tropical regions where soil processes are not so limited by climate. Carbon sequestration, for example, across the Great Plains of the United States, depends on interactions between annual precipitation and soil type (Burke et al. 1989). Nutrients can be lost due to a number of factors. For example, in tall-grass prairie, the major local pathways for nutrient loss are either abiotic (hydrologic fluxes and fire-induced losses) or biotic (gaseous fluxes via denitrification and ammonia volatilization) (Blair et al. 1998). In wet grassland systems, hydrology is indisputably the dominant controlling factor. However, in waterlogged soils, there can be considerable release of gaseous products from denitrification, while at the same time waterlogging limits nitrification and mineralization (Nijburg & Laanbroek 1997).

Forests

Of the total terrestrial net primary productivity of 62.6 Pg C yr^{-1}, forests are responsible for 52 percent (ca. 32.6 Pg C yr^{-1}) (IPCC 2001; Saugier et al. 2001; Gower 2002). When combined with the large land surface area covered by forests (boreal, temperate, and tropical), savannas, and shrubland systems (ca. 72×10^6 km^2), it is clear that nearly 50 percent of the organic input into soils are from forest litter and plant roots. The role of the forest soil biota is immediately highlighted when we realize that the majority of the world's forests rely heavily on internal nutrient cycling to maintain productivity (see, for example, Miller et al. 1979) and, in the total absence of soil biota, these systems would become extremely nutrient limited.

Not only is nutrient cycling controlled by the soil biota but several major sources of nutrient input to forests (e.g., mineral weathering and N fixation) are also mediated or strongly affected by soil biological activity. Thus, factors that impair soil biological activity (e.g., acidification, heavy metal pollution) have negative consequences for both nutrient inputs and subsequent cycling. There is also an increasing appreciation that our understanding of forest nutrient cycling and the high inorganic N concentrations currently found in many of the soils of the major developed regions of the world are atypical of unmanaged, pristine forests (Perakis & Hedin 2002), where more tightly coupled nutrient cycling, frequently involving intimate links between primary producers and soil biota, may dominate (Ohlund & Nasholm 2001). At the most basic level, forest soil biota are essential for the maintenance of the forest, but these organisms also have very direct effects on the ecosystem services provided within the forest.

Although examples exist that show where soil biodiversity is critical in maintaining an ecosystem service (e.g., specific edible fungal fruiting bodies, inoculation or manipulation of wood stump saprotrophs to control pathogens), in general we are ignorant of the extent to which soil diversity is important in maintaining services. The observation that increased inorganic N additions to forest soils can greatly impair methane oxidation rates (Wang & Ineson 2003) is an example of how soil organisms are clearly controlling an important ecosystem service, but direct information of the diversity of the species involved is proving extremely difficult to extract (Bull et al. 2000).

In unmanaged humid tropical forests it is rare that any specific identifiable group of organisms have a role in soil processes. Litter breakdown is accomplished by many disparate groups of organisms such as crabs, millipedes, cockroaches, and so on. Within these groups there is always some marginal overlap in feeding-niche parameters. Similarly, soil-feeding termites (which exhibit particularly high species diversity in African forests) produce significant methane fluxes and play an important role in P cycling and soil organic matter turnover. However, all these processes are also mediated by other animals and microbes, and strongly buffered by the biophysical pools of soil organic matter and ion exchange sites (Anderson 1994). Rate determinants of key processes, such as hydrologic pathways and erosion, are affected by the combined activities of microbes, animals, and plant roots on aggregate stability and soil structure (Silver et al. 2000).

Arable Land

Intensive tilled arable farming systems occur worldwide, but are most prominent in industrialized countries in temperate regions. Intensive tilled agriculture bears little resemblance to highly complex unmanaged ecosystems. Arable soils are extremely disturbed by cultivation, fertilizers, and pesticides, all of which alter their biophysical composition (Anderson 2000). Plant communities of these conventionally managed, tilled agroecosystems are unnaturally simple. Intensive tillage farming negates the activity of many soil organisms, such as earthworms (Marinissen & de Ruiter 1993), ants (Decaens et al. 2002), and mycorrhizal fungi (Helgason et al. 1998), but some soil organisms continue to play an important role in the decomposition of crop residuals or organic manure, and are involved in the processes of nitrification and pesticide degradation (Anderson 2000).

In intensive tilled agriculture, short rotations and the use of high-yielding crops can trigger development of soil-borne pathogens and viruses, nematode infestations, and soil-dwelling insects, causing major yield losses. Overcoming these outbreaks requires the breeding of resistant crop varieties, the frequent use of soil pesticides (Epstein & Bassein 2001), or the development of novel biological control practices (Whipps 2001). In contrast, in traditional small-holder (predominantly tropical) farms, the role of organisms is apparent and the activities of a few species of earthworms, termites, and other fauna dominate soil communities. After hand tillage (mostly on tropical farms) or abandonment of the site in shifting agriculture, soil earthworm communities usually recover rapidly (Decaens & Jimenez 2002). This is in contrast with the microbial recovery of soil communities after intensive tilled farmed systems in temperate zones (Siepel 1996; Korthals et al. 2001).

Intensively tilled arable soils are primarily controlled by abiotic factors. For example, erosion is a constraint to soil fertility in 10 to 20 percent of the world's regions (Bot et al. 2000). In parts of the tropics, seasonal rains can cause severe soil erosion, degrading soils and making agriculture less sustainable. The adoption of minimum tillage or organic farming in some cropping systems to improve soil organic matter, or the adoption of erosion control for improved soil moisture, results in a significant recovery of soil biodiversity (Mader et al. 2002). However, this recovery, including the return of earthworms, has not been consistently linked to improvement in crop yields (Brown et al. 1999; Mader et al. 2002).

Arable soils may sequester carbon, but this ecosystem service depends on a combination of climatic factors (temperature/moisture conditions), topography and soil properties (texture, clay content, mineralogy, acidity) (Robert 2001), deposition history (VandenBygaart et al. 2002), and management strategy (Brye et al. 2002).

Management Trade-Offs and Functional Diversity in Grassland, Forest, and Arable Land

In evaluating the consequences of management options for ecosystem services, it is impossible to consider all possible combinations of potentially interacting factors, such as soil type, vegetation type, and climate. Therefore specific examples were chosen, which we believe may be adopted for any type of ecosystem at any place on earth. Tabulating our information (see Tables 2.2–2.4) gives us the opportunity to quickly assess where specific soil organisms contribute positively, neutrally, or negatively (relative to abiotic factors) to the sustainable delivery of specific ecosystem goods or services.

In the specific sections for grassland, forest and arable land, we discuss trade-offs for management versus no (or extensive) management. In the case of arable land, we compare tilled land with non-tilled land. There is little consistency across studies with regard to the role of species diversity within functional groups of soil organisms on ecosystem processes (Mikola et al. 2002). However, our method of representation enables us to draw some general conclusions about the importance of functional diversity of soil organisms for the sustainable delivery of ecosystem goods and services in the three major types of ecosystems considered.

Temperate systems were selected because they are the best characterized. We appreciate that the assessment is imperfect, as perceptions of the services provided by each system in a locale differ and we therefore cannot list and evaluate precisely the same services in each ecosystem. We have compared the extremes of unmanaged and managed systems to highlight the service trade-offs, which are implicitly brought about by management. In reality, there is usually a gradient of possible interventions.

The relative importance of ecosystem goods and services ranking from −3 (strongly negative) to +3 (strongly positive) under both unmanaged (pristine, semi-natural, or zero-input) and managed (with significant human inputs and/or anthropogenic disturbance) states are presented. We compare three specimen temperate ecosystems: grassland, forest, and arable land. Under each category we have distinguished between biotic (mediated by living organisms) and abiotic (mediated by chemical, physical, and geologically historical factors, over which living organisms have, essentially, no overwhelming control on the short- and medium-term) processes. This is done by allocating each an asterisk rating from * (weak) to *** (strong) relative importance in determining the impact of the biotic and abiotic process on the specific ecosystem good or service identified. Habitat support functions, such as decomposition, bioturbation, and so on, are included in the allocation of drivers to biotic and abiotic categories. This level of resolution indicates those systems in which the manipulation of soil biodiversity (within or between groups of soil organisms that perform specific functions) could be effective in reinforcing a particular service, and where the services appear most vulnerable to changes in the biological soil community (see Wardle et al., Chapter 5, where some processes have been valued slightly different owing to the comparison between non-perturbed and perturbed systems).

Grasslands

Unmanaged and lightly managed grasslands can be seen as having a low rank for ecosystem service for products directly consumed by human beings, but they often sustain populations of large herbivores that serve as food and other animal products used by humans (Table 2.2). Unmanaged grasslands are extensive in the tropics, and range along a broad gradient of precipitation that strongly influences both production and biological diversity, aboveground and belowground, merging into parkland and, ultimately, forest at the wet end. These grasslands are often the guardians of important watersheds. They have a disproportionately high aesthetic value, deriving from an inherently high biodiversity, the visibility of some wildlife, and a large recreational potential (McNaughton et al. 1983). Where grasslands are the product of low precipitation, management options are limited because animal production is ultimately a product of (inverse) stocking density (Ruess & McNaughton 1987). However, in mesic systems, where many managed grasslands are the product of historical forest clearance and/or intensive grazing, production can often be enhanced in proportion to fertilizer input and manipulation (through tillage and seeding) of plant species composition. Such management carries major implications for water quality (reduced by large rises in dissolved C and N compounds in runoff water) and for greenhouse gas emissions (Williams et al. 1998). There is also, arguably, a fall in aesthetic value and restrictions in the availability of land for recreational purposes.

A large functional diversity of soil organisms in grassland soils contributes to the sustainability of ecosystem goods and services. In unmanaged grasslands, there is a strong positive role of soil organisms, whereas intensified grassland management results in reducing the role of soil organisms in the decomposition of soil organic matter and the mineralization of nutrients due to the addition of mineral fertilizers or liquid manure (Bardgett & Cook 1998). Reduced activity of burrowing soil organisms decreases the contribution of managed grasslands to water storage, which results in enhanced runoff of rainwater and, consequently, flooding in downstream areas (Bardgett et al. 2001). In managed grasslands, nitrifying microorganisms may contribute to the leaching of mineral nitrogen to surface water or groundwater because the amount of available mineral nitrogen exceeds the uptake by plants (Smith et al. 2002). The predominance of bacterial-based soil food webs reduces the capacity of managed grasslands to act as sinks for carbon (Burke et al. 1989; Brye et al. 2002; Mader et al. 2002).

The possible management interventions are mowing, stocking density, fertilization, pesticide application, seeding, tillage, and enclosure. In reality, there is a gradient of management from light to intensive, and also a major difference between dryland grassland systems and mesic ones. The total number of land uses imposed on managed grasslands is very large. Grasslands may also be created on a temporary basis when tropical forest is felled, or permanently by invasive species if the soil is subsequently exhausted

Table 2.2. Provision of goods and services in a temperate grassland ecosystem.

This ecosystem is considered in its unmanaged and managed states. A rank from −3 (strong disservice) through 0 (neutral) to +3 (strong service) is given for each good or service, indicating its value to human societies. Under each category we have distinguished between biotic (mediated by living soil organisms) and abiotic (mediated by chemical, physical and geological, climatological or historical factors, over which living organisms have, essentially, no overwhelming control on the short and medium term) processes. The relative contributions of biotic and abiotic processes (= ecosystem functions, including anthropogenic inputs and perturbations) to each good or service is given by asterisks * (small) to ***(large). Note that in Chapter 5, Tables 5.A1–5.A3 have been developed using the same principle, but there the relative contributions of abiotic and biotic processes have sometimes been slightly differently valued. This is due to the comparison between unmanaged and managed.

Goods or Services	Unmanaged Grassland #			Managed Grassland ##		
	Rank	Biotic	Abiotic	Rank	Biotic	Abiotic
Food production Plant products	0			0		
Animal products	2	*** decomposition, bioturbation, organic matter transformation + nutrient cycling, resistance to pests & diseases	** soil type, topography, fire, climate	3	* nutrient transformation, nitrification, etc.	*** chemical inputs, soil type, topography, climate
Water quality (Riparian context: quality to local streams and rivers)	2	*** retention of N in biomass, physical stabilization, interception of run-off, soil organisms	** soil type and cover, topography, fire, climate (esp. precipitation)	−3	*** nitrification, Phosphate liberation, DON, DOC, microbiological pollution of runoff, soil organisms	*** soil type and cover, topography, fire, climate (esp. precipitation), compaction
Water volume† (Flow to watersheds)	2	*** moisture retention by OM, evapotranspiration, soil organisms	** soil type and cover, topography, fire, climate (esp. precipitation), infiltration, runoff	−2.	* moisture retention by organic matter, soil organisms	*** soil type and cover, topography, fire, climate (esp. precipitation), flood, run-off, infiltration, compaction

(continued)

Table 2.2. (continued)

Goods or Services	Unmanaged Grassland #			Managed Grassland ##		
	Rank	Biotic	Abiotic	Rank	Biotic	Abiotic
Fiber (e.g., wool, leather)	2	*** decomposition, bioturbation, organic matter transformation + nutrient cycling, resistance to pests & diseases	*** soil type, topography fire, climate	1	* nutrient transformation, e.g., nitrification etc.	*** chemical inputs, soil type, topography, climate
Recreation	2	*** wildlife	** landscape	1	* wildlife, hiking, etc., according to habitat	*** amenity value reduction, e.g., topography amendment, fencing, denial of access
C sequestration (storage of carbon in biomass or soil organic matter, mitigating global warming)	2‡	*** organic matter formation/accumulation, CaCO$_3$ deposition	** complexing organic matter, texture, fire, CaCO$_3$ deposition	1	** net C accumulation (despite bacterial based foodwebs enhancing C loss)	** complexing organic matter, texture, fire, CaCO$_3$ deposition
Trace gases and atmospheric regulation (production of CH$_4$ and Noxides by microbes, also oxidation of CH$_4$)	2	*** maintenance of C and N balances	** texture, climate, pH	−3	*** nitrifiers, denitrifiers, loss of CH$_4$ oxidation	** texture, climate, pH

includes lightly managed freerange grasslands with no fertilizer and pesticide inputs, excluding winter livestock feed.

intensively managed animal and plant production systems with fertilizer and pesticide inputs and/or irrigation.

† not including charging of groundwater.

‡ because of higher accumulation of surface and subsurface organic matter.

through subsistence agriculture. Unmanaged grasslands are mainly used for free-range animal production.

In conclusion, our example shows that in the type of grasslands considered, the largest impacts of management on ecosystem goods and services are on water quality and quantity and on trace gases. There may be a management trade-off between enhanced production and increased leaching and trace gas production, but in this trade-off the total area needed for food production as well as the price of land will undoubtedly play an important role.

Forests

Unmanaged forest ecosystems provide a diversity of ecosystem goods and services, and their ranks are moderate (food production) to strongly positive (e.g., fuel/energy, recreation, carbon sequestration, and regulating trace gases) (Table 2.3). Biotic and abiotic processes have almost equal contributions to the rate or efficiency of the delivery of these goods and services. Plant roots play an important role in the regulation of the quality and quantity of water volume and erosion control, whereas soil bacteria are particularly important for the regulation of the trace gases in the atmosphere (Wang & Ineson 2003). Mycorrhizal fungi play an important role in the formation of organic matter in coniferous forests (Smith & Read 1997), whereas earthworms have some impact in deciduous forests (Lavelle et al. 1997).

In managed forests, the presence of soil pathogens, parasites, decomposers, and N_2 fixers is a more important factor, especially in monospecific stands for timber or fuelwood production (Waring et al. 1987). On the other hand, there are fewer natural sources of food, a less diverse range of macro- and microorganisms, and fewer soil habitats in managed forests. Conversely, the importance of abiotic factors in controlling the delivery of these ecosystem goods and services is little changed in managed forests, except for the potential role of external inputs, such as fertilizers and lime, in altering trace gas production, mineralization, and nitrate leaching.

Intensifying forest management decreases the potential for carbon sequestration into the soil, and it also reduces sources of biochemicals and medicines by providing fewer habitats for other organisms. There is considerable diversity of functions of soil organisms in unmanaged and managed forests. Soil fungi play a more substantial role in decomposition in forests than in grasslands, and the role of plant roots in the delivery of ecosystem goods and services, for example in erosion control, is also more prominent. Forest management has relatively little effect on functional diversity of soil organisms, but it may change the importance of symbiotic nitrogen-fixing microorganisms to nitrogen availability, especially when monocultures of nitrogen-fixing trees are established. Humans can affect the diversity of mycorrhizal fungi, as can atmospheric deposition (Smith & Read 1997).

In conclusion, trade-offs for forest management occur within narrower margins

Table 2.3 Provision of goods and services in temperate unmanaged and managed forest ecosystems.

For further explanations, see Table 2.2.

Goods or Services	Unmanaged Forest			Managed Forest		
	Rank	Biotic	Abiotic	Rank	Biotic	Abiotic
Food production	1	*** decomposition, fungal diversity	* soil type, pH	0		
Water quality and volume; flood and erosion control	3	*** evapotranspiration *** plant composition ** hydraulic lift *** rooting depth * soil organisms, bioturbation	*** shading/cover *** stabilization *** runoff, infiltration *** slope, topography, soil physical properties, soil type	3	*** evapotranspiration *** plant composition ** hydraulic lift *** rooting depth * soil organisms, bioturbation	*** shading/cover *** stabilization *** runoff, infiltration *** slope, topography, soil physical properties, soil type
Fiber	2	** net primary production	*** soil type, climate	3	*** net primary production	** soil type, climate *** chemical inputs
Fuel/Energy	3	** net primary production	*** soil type, climate	3	** net primary production	** soil type, climate *** chemical inputs
Biochemicals and medicines	3	decomposition ** microbial diversity		0		
Habitat provision	3	*** intrinsic biodiversity, NPP, ecosystem engineers	*** soil type, topography, climate	2	** intrinsic biodiversity, NPP, ecosystem engineers	*** soil type, topography, climate

Service	Score			Score		
Waste disposal	0			0		
Biological control	2	*** intrinsic biodiversity	** soil type and properties	1	** intrinsic biodiversity	* soil type and properties
Recreation	3	*** wildlife	** landscape	2	*** wildlife	** landscape
C sequestration Deciduous	3	organic matter for-mation/accumulation * litter quality * earthworms *** roots	* texture, climate, CaCO$_3$ deposition	2	organic matter for-mation/accumulation * litter quality * earthworms * roots	* texture, climate, CaCO$_3$ deposition
Coniferous (NB: no worms)	2	organic matter formation ** litter quality * roots **/*** mycorrhiza * pathogens	** texture, climate ** fire CaCO$_3$ deposition	1	** litter quality * roots ** pathogens **/*** mycorrhiza	** texture, climate */** fire, CaCO$_3$ deposition
Trace gases and atmospheric regulation	3	*** maintenance of C and N balances	** texture, climate, pH	3	*** nitrifiers, denitrifiers, CH$_4$ oxidation	** texture, climate, pH

than they do for grasslands, as most options differ by a unit of only one. However, forest management has large implications for biodiversity within forests, for example, the removal of wood, and/or soil erosion following large-scale tree felling (see Ineson et al., Chapter 9), can have important impacts on soil biodiversity and soil ecosystem services. Management for biodiversity and ecosystem functioning will have to balance the trade-offs of selective tree cutting against felling of entire forest stands (see Ineson et al., Chapter 9). Selective felling of trees leaves a large proportion of the forest system intact, while complete felling can result in complete loss of topsoil due to erosion. Moreover, selective forest management promotes more attractive forests for recreational use.

Arable Land

Agriculture ranges in management from non-tilled, or lightly managed, lightly cultivated fields and food-gathering systems to highly managed tilled fields that receive large inputs of pesticides, fertilizers, energy in tillage, and even irrigation water. Another type of distinction is conventional versus biological (or organic) agriculture. We choose soil tillage involving heavy soil disturbance as an example, which has a major impact on soil organisms (Marinissen & de Ruiter 1993; Helgason et al. 1998). It has been demonstrated that high yields can be obtained with monocropping and intensive tillage management, where natural biodiversity is overridden by high inputs of nutrients and pesticides (Tilman et al. 2002). This, however, often comes with high costs to the environment in water pollution by nitrates and toxic, partially decomposed, chemical inputs (Carpenter et al. 1998). It also can result in fluxes of CO_2 from fossil fuel combustion and N_2O (Hall et al. 1996), one of the most active greenhouse gases (Table 2.4).

Biodiversity, or the variety of life forms, does not necessarily increase the sustainable delivery of ecosystem goods and services in agriculture. Nitrifiers and denitrifiers are responsible for huge losses of nitrogen from the soil system while leading to pollution of the groundwater with NO_3 and the atmosphere with N_2O (van Breemen et al. 2002). Many closely coupled natural systems operate essentially without these organisms in that they have an NH_4-N nutritional system (Coleman & Crossley 1996). Nitrogen fixers are essential for food production, especially in farming systems where mineral fertilizers are costly and organic fertilizers are insufficiently available. Soybeans are considered a desirable crop because of their N-fixation potential that supplants the need for N fertilization. They, however, produce little crop residue that is easily degraded and thus are not desirable for carbon sequestration unless grown in a cropping sequence with a high residue producer such as maize, unless the maize shoots are used as fuel.

On average, most systems that have and produce soil biodiversity also favor carbon sequestration, good soil tilth, and high fertility (Sperow et al. 2003). This can occur in non-tillage systems and in systems with increased cropping complexity and perennial crops, as well as in systems where nutrient inputs are efficiently utilized. Site-specific

Table 2.4 Provision of goods and services in non-tilled and tilled temperate arable land ecosystems.

For further explanations, see Table 2.2.

Goods or Services	Non-Tilled Arable			Tilled Arable		
	Rank	Biotic	Abiotic	Rank	Biotic	Abiotic
Food production (plant + animal)	3	* plant breeding (roots and residues), bioturbation	*** climate, chemical inputs	3	*** plant breeding (roots and residues)	* climate, tillage, chemical inputs
Water quality	−1	* nitrification, pesticide degradation, leaching by biopores	** topography, climate, infiltration, runoff, soil type and cover	−2	* nitrification, pesticide degradation and mobilization	*** topography, climate, infiltration, runoff, soil type and cover, leaching, volatilization of NH_4
Water volume	−1	** moisture retention by OM, evapo-transpiration, biopores by soil organisms	** topography, climate, run-off, soil type, cover		* moisture retention by OM, higher evapotranspiration and less biopores by soil organisms	*** topography, climate, run-off, soil type
fiber	3	* plant breeding (roots and residues), bioturbation	*** climate, chemical inputs	3	*** plant breeding (roots and residues)	* climate, tillage, chemical inputs
Waste disposal	1	** decomposition, co-metabolism, build-up of inter-mediate toxic products	* volatilization, water regime, soil properties, sequestration, incorporation by machinery	3	** decomposition, co-metabolism, build-up of inter-mediate toxic products	** volatilization, water regime, soil properties, sequestration, incorpor-ation by machinery

(continued)

Table 2.4 *(continued)*

Goods or Services	Non-Tilled Arable			Tilled Arable		
	Rank	Biotic	Abiotic	Rank	Biotic	Abiotic
Biological control	−1	** crop rotation, predators, bacterial and fungal antagonists, mycorrhizal fungi, GMOs	* pesticides	0–1	** crop rotation, predators, bacterial and fungal antagonists, mycorrhizal fungi, GMOs	** pesticides, soil tillage
Recreation	1–2	** choice of crop species, composition of field margins	** landscape	1–2	** choice of crop species; composition of field margins	** landscape
C sequestration	0–1	* decomposition	* texture, nature of clay minerals, climate	−1–0	* decomposition	** topography, texture, nature of clay minerals, soil structure
Trace gases and atmospheric regulation	−1	** ammonification, nitrification, fungal decomposition	** moisture regime, soil structure	−3	** ammonification, nitrification, bacterial decomposition	** moisture regime, soil structure

exceptions require careful management. Non-tillage agriculture, when used with certain soils, increases the incidence of plant diseases. It can also lead to wet cold soils that delay planting. Increased cropping complexity and cover crops can have significant advantages in many agroecosystems, especially where the degree of mechanization is relatively low. On some soils, however, the moisture used by the cover crop results in lower crop yields and if not properly managed some cover crops can compete with the primary crop and act as weeds (Locke et al. 2002).

In conclusion, management trade-offs for arable land, for example non-tillage versus tillage, need to balance between the efficiency of fertilizer inputs and obtaining high crop yields. Indirectly, the price of land and labor will be weighed against that of fertilizer, pesticides, and the costs related to the resulting environmental pollution, such as groundwater contamination. Non-tillage favors soil C sequestration, but can enhance the need for disease control. When comparing conventional (highly intensive) agriculture with organic (or biological) agriculture, similar trade-offs may occur, possibly with a stronger emphasis on reduced outputs versus the price of land.

Discussion and Conclusions

Soil organisms play a major role in the delivery of ecosystem goods and services that are crucial for supporting human societies and for the sustainability of natural and managed ecosystems. Soil organisms act on very small scales, but their effects may range from local (diseased plants, nutrient mineralization) to very large scales (plant succession, carbon sequestration, production of trace gases that contribute to global warming). The diversity of soil organisms may matter more for a process that is accounted for by only a few species than for a process that is accounted for by many species. However, empirical evidence on effects of species diversity on ecosystem processes is still relatively rare and does not yet allow generalizations. It is probable that the diversity of functions is more important for the sustainability of ecosystem goods and services than species diversity *per se,* but this area is still wide open for further studies.

The role of soil organisms is more prominent in grasslands (especially natural ones), forests, and low-input (no-till) arable land than in intensively managed grasslands and arable land. However, the relative importance of soil organisms for the performance of ecosystem processes (as compared with the importance of abiotic influences) differs along climatic gradients or between soil types. There are also differences between the relative contributions of different taxa of soil organisms to ecosystem processes along climate gradients or between soil or vegetation types. In cold areas, for example, soil microorganisms play a lesser role in the decomposition of organic matter, whereas soil fauna have a more dominant role. Earthworms are key species in mesic grasslands, but enchytraeids are crucial in coniferous forests and some arable land.

Human interventions, such as plowing, fertilization, and using pesticides, often lead to shifts in the major decomposition channels or to a by-passing of the role of soil organisms. In intensively fertilized tilled arable land, the decomposition pathway is bacteria-based and the role of symbiotic mutualists (mycorrhizal fungi and nitrogen-fixing microorganisms) is largely redundant. Stability of nutrient pools in these systems may be achieved by high-input measures, but this results in, for example, the leakage of nutrients to ground- and surface water. In these cases, human activity to enhance the delivery of ecosystem goods, such as food production, result in the loss of ecosystem services, such as water purification occurring in soils.

Other human-induced changes, such as land use change, deforestation, soil drainage, erosion, enhanced temperature, and increased CO_2 concentrations may all affect soil ecosystem goods and services by affecting soil organisms either directly or indirectly. Erosion will affect soil communities through the direct loss of habitat, whereas rising temperature and CO_2 concentration may lead to more incipient changes in soil communities and, therefore, of the functioning of soil systems and the sustainability of the delivery of ecosystem goods and services.

There are clearly management trade-offs for the role of soil organisms in the delivery of ecosystem goods and services. We do not have much evidence that these trade-offs act through the loss of species diversity, but this is mainly due to our limited knowledge on the diversity of, for example, soil microbial communities and consequences for ecosystem processes. Management trade-offs clearly act through effects on the diversity of functions. Intensive tillage farming practice reduces the abundance of earthworms, which negatively affects both the water-holding capacity of soils as well as the population of mycorrhizal fungi. This, as well as changes in the soil due to deforestation, may enhance flooding incidence in lowlands due to increased peaks in run-off water.

The template that we have developed for the analysis of the contribution of soil organisms and abiotic soil factors to the organization of soil processes, and to the delivery of ecosystem goods and services, is applicable to a wide range of environmental contexts. The examples that we have presented, however, apply to temperate systems. The approach adopted here may well prove valuable for comparison with tropical systems, where the potential for soil biotic diversity may be higher.

The obvious weakness of the present approach is that the relative importance of the services and the relative contributions of biotic (essentially manageable) and abiotic (only partially manageable) processes are expressed only in comparative terms. To make absolute (monetary) valuations possible, some of the services (e.g., food and fuel production) could be costed for a given local economy, while other services could be assigned financial status from these by reference to the relative importance we have suggested. The difference between these biological valuations and other schemes of costing, which are the stock-in-trade of economists, is that in each defined ecosystem some processes are amenable to management, and others are not. The prices of services, which essentially reflect their availability for manipulation by humans, should be adjusted accordingly.

Research Needs and Recommendations

Soil biota provide many services in a wide range of terrestrial ecosystems. Our knowledge of how to manage and protect species in the soil and the processes that they drive is, however, limited (Wall et al. 2001). Areas that need further studies in order to enhance the effectiveness of management are:

1. *Incorporate the role of soil organisms and soil biodiversity in crop protection.* Soil organisms influencing plant defense against aboveground insects and pathogens and soil management may, therefore, influence plant protection in arable ecosystems.
2. *Acknowledge the role of soil biota in restoration and conservation of aboveground biodiversity.* Soil organisms are strongly involved in primary and secondary succession and in the regulation of plant species diversity in unmanaged ecosystems. More studies are needed to determine how these processes operate and how they can be used and influenced in order to reach management goals, such the conversion from arable land to more natural systems in order to conserve and protect biodiversity.
3. *Use soil organisms in bioremediation.* Many soil organisms can play a role in cleaning polluted soils and the sheer diversity of microbes provides ample opportunities for reducing pollution loads in contaminated soils.
4. *Use food web modeling to improve the conservation and use of soil nutrients.* Soil ecology has been strong in developing functional group approaches and in modeling the interactions between functional groups in order to assess the stability of ecosystem processes. These food web models may be further developed for use in testing land management options—for example, in relation to land use history, current status of the soil abiotic and biotic conditions, and management goals.
5. *Communicate the role of soil biota and soil biodiversity to land managers and policy makers.* Soil organisms for too long have been "out of sight, out of mind." However, increasingly, land managers and policy makers express interest in the sheer diversity underneath their feet. Communication of the relation between soil biota, soil biodiversity, ecosystem processes, and ecosystem services and practical recommendations are, therefore, of top priority. We hope that this chapter will inspire end users and stakeholders to start collaborative actions leading to enhancing both knowledge about soil biodiversity and ecosystem functioning and the application of these results in order to improve the sustainability of ecosystem goods and services as provided by the soil biota.

Literature Cited

Anderson, J.M. 1994. Functional attributes of biodiversity in land use systems. In: *Soil Resilience and Sustainable Land Use*, edited by D.J. Greenland and I. Szabolcs, pp. 267–290. Wallingford, UK, CAB International.

Anderson, J.M. 2000. Food web functioning and ecosystem processes: Problems and perceptions of scaling. In: *Invertebrates as Webmasters in Ecosystems*, edited by D.C. Coleman and P.F. Hendrix, pp. 3–24. Wallingford, UK, CAB International.

Baas, W.J. 1989. Secondary plant compounds, their ecological significance and consequences for the carbon budget. Introduction of the carbon/nutrient cycle theory. In: *Causes and Consequences of Variation in Growth Rate and Productivity of Higher Plants*, edited by H. Lambers, pp. 313–340. The Hague, SPB Academic Publishing.

Bardgett, R.D., J.M. Anderson, V.M. Behan-Pelletier, L. Brussaard, D.C. Coleman, C. Ettema, A. Moldenke, S.P. Schimel, and D.H. Wall. 2001. The role of soil biodiversity in the transfer of materials between terrestrial and aquatic systems. *Ecosystems* 4:421–429.

Bardgett, R.D., and R. Cook. 1998. Functional aspects of soil animal diversity in agricultural grasslands. *Applied Soil Ecology* 10:263–276.

Bardgett, R.D., and D.A. Wardle. 2003. Herbivore mediated linkages between aboveground and belowground communities. *Ecology* 84:2258–2268.

Bever, J.D. 1994. Feedback between plants and their soil communities in an old field community. *Ecology* 75:1965–1977.

Blair, J.M., T.R. Seastedt, C.W. Rice, and R.A. Ramundo. 1998. Terrestrial nutrient cycling in a tallgrass prairie. In: *Grassland Dynamics*, edited by A.K. Knapp, J.M. Briggs, D.C. Hartnett, and S.L. Collins, pp. 222–243. Oxford, Oxford University Press.

Bot, A.J., F.O. Nachtergaele, and A. Young. 2000. Land resource potential and constraints at regional and country levels. World Soil Resources Report 90. FAO Land and Water Development Division, 114 pp. ftp://ftp.fao.org/agl/agll/docs/wsr.pdf.

Brown, G.G., B. Pashanasi, C. Villenave, J.C. Patrón, B. Senapati, S. Giri, I. Barois, P. Lavelle, E. Blanchart, R.J. Blakemore, A. Spain, and J. Boyer. 1999. Effects of earthworms on plant production in the tropics. In: *Earthworm Management in Tropical Agroecosystems*, edited by P. Lavelle, L. Brussaard, and P.F. Hendrix, pp. 87–148. Wallingford, UK, CAB International.

Brown, V.K., and A.C. Gange. 1989. Herbivory by soil-dwelling insects depresses plant species richness. *Functional Ecology* 3:667–671.

Brown, V.K., and A.C. Gange. 1992. Secondary plant succession—how is it modified by insect herbivory. *Vegetatio* 101:3–13.

Brye, K.R., T.S. Gower, J.M. Norman, and L.G. Bundy. 2002. Carbon budgets for a prairie and agroecosystems: Effects of land use and interannual variability. *Ecological Applications* 12:962–979.

Bull, I.D., N.R. Parekh, G.H. Hall, P. Ineson, and R.E. Evershed. 2000. Detection and classification of atmospheric methane oxidizing bacteria in soil. *Nature* 405:175–178.

Burke, I.C., C.M. Yonker, W.J. Parton, C.V. Cole, K. Flach and D.S. Schimel. 1989. Texture, climate, and cultivation effects on soil organic matter content in U.S. grassland soils. *Soil Science Society of America Journal* 53:800–805.

Carpenter, S.R., N.F. Caraco, D.L. Correll, R.W. Howarth, A.N. Sharpley, and V.H. Smith. 1998. Nonpoint pollution of surface waters with phosphorus and nitrogen. *Ecological Applications* 8:559–568.

Coleman, D.C., and D.A. Crossley. 1996. *Fundamentals of Soil Ecology*. San Diego, California, Academic Press.

Daily, G.C., S.E. Alexander, P.R. Ehrlich, L.H. Goulder, J. Lubchenco, P.A. Matson, H.A. Mooney, S. Postel, S.H. Schneider, D. Tilman, and G.M. Woodwell. 1997.

Ecosystem services: Benefits supplied to human societies by natural ecosystems. *Issues in Ecology* 2:1–18.

Decaens, T.N., N. Asakawa, J.H. Galvis, R.J. Thomas, and E. Amezquita. 2002. Surface activity of soil ecosystem engineers and soil structure in contrasted land use systems of Colombia. *European Journal of Soil Biology* 38:267–271.

Decaens, T.N., and J.J. Jimenez. 2002. Earthworm communities under an agricultural intensification gradient in Colombia. *Plant and Soil* 240:133–143.

De Deyn, G.B., C.E. Raaijmakers, H.R. Zoomer, M.P. Berg, P.C. de Ruiter, H.A. Verhoef, T.M. Bezemer, and W.H. van der Putten. 2003. Soil invertebrate fauna enhances grassland succession and diversity. *Nature* 422:711–713.

Epstein, L., and S. Bassein. 2001. Pesticide applications of copper on perennial crops in California, 1993 to 1998. *Journal of Environmental Quality* 30:1844–1847.

Gonzalez, G., and T.R. Seastedt. 2001. Soil fauna and plant litter decomposition in tropical and subalpine forests. *Ecology* 82:955–964.

Gower, S.T. 2002. Productivity of Terrestrial Ecosystems. In: *Encyclopedia of Global Change, Vol. 2,* edited by H.A. Mooney and J. Canadell, pp. 516–521. Oxford, Blackwell Scientific.

Griffiths, B.S., K. Ritz, R.D. Bardgett, R. Cook, S. Christensen, F. Ekelund, S.J. Sørenson, E. Bååth, J. Bloem, P.C. de Ruiter, J. Dolfing, and B. Nicolardot. 2000. Ecosystem response of pasture soil communities to fumigation-induced microbial diversity reductions: An examination of the biodiversity-ecosystem function relationship. *Oikos* 90:279–294.

Grime, J.P. 1979. *Plant Strategies and Vegetation Processes.* Chichester, UK, John Wiley & Sons.

Hall, S.J., P.A. Matson, and P. Roth. 1996. NO_2 emission from soil: Implications for air quality modelling in agricultural regions. *Annual Review of Energy and Environment* 21:311–346.

Helgason, T., T.J. Daniell, R. Husband, A.H. Fitter, and J.P.Y. Young. 1998. Ploughing up the wood-wide web? *Nature* 394:431.

Hooper, D.U., and P.M. Vitousek. 1998. Effects of plant composition and diversity on nutrient cycling. *Ecological Monographs* 68:121–149.

IPCC. 2001. Intergovernmental Panel on Climate Change. *Climate Change 2001: The Scientific Basis.* Intergovernmental Panel on Climate Change, Cambridge, UK, Cambridge University Press.

Klironomos, J.N. 2002. Feedback to the soil community contributes to plant rarity and invasiveness in communities. *Nature* 417:67–70.

Korthals, G.W., P. Smilauer, C. Van Dijk, and W.H. van der Putten. 2001. Linking above- and belowground biodiversity: Abundance and trophic complexity in soil as a response to experimental plant communities on abandoned arable land. *Functional Ecology* 15:506–514.

Lavelle, P. 1997. Faunal activities and soil processes: adaptive strategies that determine ecosystem function. *Advances in Ecological Research* 27:93–132.

Lavelle, P. 2002. Functional domains in soils. *Ecological Research* 17:441–450.

Lavelle, P., D. Bignell, M. Lepage, V. Wolters, P. Roger, P. Ineson, O.W. Heal, and S. Dhillion. 1997. Soil function in a changing world: The role of invertebrate ecosystem engineers. *European Journal of Soil Biology* 33:159–193.

Lavelle, P., E. Blanchart, A. Martin, S. Martin, A. Spain, F. Toutain, I. Barois, and R.

Schaefer. 1993. A hierarchical model for decomposition in terrestrial ecosystems: Application to soils of the humid tropics. *Biotropica* 25:130–150.

Locke, M.A., K.N. Reddy, and R.M. Zablotowicz. 2002. Weed management in conservation crop production systems. *Weed Biology and Management* 2:123–132.

Mader, P., A. Fliessback, D. Dubois, L. Gunst, P. Fried, U. Niggli. 2002. Organic farming and energy efficiency. *Science* 298:1891.

Marinissen, J.C.Y., and P.C. de Ruiter. 1993. Contribution of earthworms to carbon and nitrogen cycling in agro-ecosystems. *Agriculture Ecosystems and Environment* 47:59–74.

McNaughton, S.J. 1983. Serengeti grassland ecology: The role of composite environmental factors and contingency in community organization. *Ecological Monographs* 53:291–320.

Mikola, J., R.D. Bardgett, and K. Hedlund. 2002. Biodiversity, ecosystem functioning, and soil decomposer food webs. In: *Biodiversity and Ecosystem Functioning: Synthesis and Perspectives*, edited by M. Loreau, S. Naeem, and P. Inchausti, pp. 169–180. Oxford, Oxford University Press.

Millennium Ecosystem Assessment. 2003. *Exosystems and Human Well-Being: A Framework for Assessment.* World Resources Institute, Washington, DC, Island Press.

Miller, H.G., J.M. Cooper, J.D. Miller, and O.J.L. Pauline. 1979. Nutrient cycles in pine and their adaptation to poor soils. *Canadian Journal of Forest Research* 9:19–26.

Nijburg, J.W., and H.J. Laanbroek. 1997. The fate of N^{15} nitrate in healthy and declining *Phragmites australis* stands. *Microbial Ecology* 34:254–262.

Northup, R.R., Z.S. Yu, R.A. Dahlgren, and K.A. Vogt. 1995. Polyphenol control of nitrogen release from pine litter. *Nature* 377:227–229.

Ohlund, J., and T. Nasholm. 2001. Growth of conifer seedlings on organic and inorganic nitrogen sources. *Tree Physiology* 21:1319–1326.

Packer, A., and K. Clay. 2000. Soil pathogens and spatial patterns of seedling mortality in a temperate tree. *Nature* 404:278–281.

Perakis, S.S., and L.O. Hedin. 2002. Nitrogen loss from unpolluted South American forests mainly via dissolved organic compounds. *Nature* 415:416–419.

Pimentel, D., C. Wilson, C. McCullum, R. Huang, P. Dwen, J. Flack, Q. Tran, T. Saltman, and B. Cliff. 1997. Economic and environmental benefits of biodiversity. *BioScience* 47:747–757.

Reinhart, K.O., A. Packer, W.H. van der Putten, and K. Clay. 2003. Plant-soil biota interactions and spatial distribution of black cherry in its native and invasive ranges. *Ecology Letters* 6:1046–1050.

Robert, M. 2001. *Soil Carbon Sequestration for Improved Land Management.* World Soil Resources Report 96. FAO Land and Water Development Division.

Ruess, R.W., and S.J. McNaughton. 1987. Grazing and the dynamics of nutrient and energy regulated microbial processes in the Serengeti grasslands. *Oikos* 49:101–110.

Saugier, B., J. Roy, and H.A. Mooney. 2001 Estimations of global terrestrial productivity: Converging toward a single number? In: *Terrestrial Global Productivity,* edited by J. Roy, B. Saugier, and H.A. Mooney, pp. 543–557. San Diego, California, Academic Press.

Siepel, H. 1996. Biodiversity of soil microarthropods: The filtering of species. *Biodiversity and Conservation* 5:251–260.

Silver, W.L., J. Neff, M. McGroddy, E. Veldkamp, M. Keller, and R. Cosme. 2000. Effects of soil texture on belowground carbon and nutrient storage in a lowland Amazonian forest ecosystem. *Ecosystems* 3:193–209.

Smith, K.A., C.P. Beckwith, A.G. Chalmers, and D.R. Jackson. 2002. Nitrate leaching following autumn and winter application of animal manures to grassland. *Soil Use and Management* 18:428–434.

Smith, S.E., and D.J. Read. 1997. *Mycorrhizal Symbiosis*. London, Academic Press.

Sperow, M., M. Eve, and K. Paustian. 2003. Potential soil C sequestration on U.S. agricultural soils. *Climatic Change* 57:319–339.

Stanton, N.L. 1988. The underground in grasslands. *Annual Review of Ecology and Systematics* 19:573–589.

Swift, M.J., O.W. Heal, and J.M. Anderson. 1979. *Decomposition in Terrestrial Ecosystems*. Oxford, Blackwell.

Tilman, D., K.G. Cassman, P.A. Matson, R. Naylor, and S. Polasky. 2002. Agricultural sustainability and intensive production practices. *Nature* 418:671–677.

van Breemen, N., E.W. Boyer, C.L. Goodale, N.A. Jaworski, K. Paustian, S.P. Seitzinger, K. Lajtha, B. Mayer, D. Van Dam, R.W. Howarth, K.J. Nadelhoffer, M. Eve, and G. Billen. 2002. Where did all the nitrogen go? Fate of nitrogen inputs to large watersheds in the Northeastern U.S.A. *Biogeochemistry* 57:267–293.

VandenBygaart, A.J., X.M. Yang, B.D. Kay, and J.D. Aspinall. 2002. Variability in carbon sequestration potential in no-till soil landscapes of southern Ontario. *Soil and Tillage Research* 65:231–241.

van der Putten, W.H. 2003. Plant defense belowground and spatio-temporal processes in natural vegetation. *Ecology* 84:2269–2280.

van der Putten, W.H., C. Van Dijk, and B.A.M. Peters. 1993. Plant-specific soil-borne diseases contribute to succession in foredune vegetation. *Nature* 362:53–56.

Wall, D.H., P.V.R Snelgrove, and A.P. Covich. 2001. Conservation priorities for soil and sediment invertebrates. In: *Conservation Biology*, edited by M.E. Soulé and G.H. Orians, pp. 99–123. Washington, DC, Island Press.

Wang, Z.P., and P. Ineson. 2003. Methane oxidation in a temperate coniferous forest soil: Effects of inorganic N. *Soil Biology and Biochemistry* 35:427–433.

Wardle, D.A. 2002. *Communities and Ecosystems: Linking the Aboveground and Belowground Components*. Princeton, Princeton University Press.

Wardle, D.A., R.D. Bardgett, J.N. Klironomos, H. Setälä, W.H. van der Putten, and D.H. Wall. 2004. Ecological linkages between aboveground and belowground biota. *Science* 304:1629–1633.

Wardle, D.A., G.W. Yeates, R.N. Watson, and K.S. Nicholson. 1995. The detritus food-web and the diversity of soil fauna as indicators of disturbance regimes in agroecosystems. *Plant and Soil* 170: 35–43.

Waring, R.H., K. Cromack Jr., P.A. Matson, R.D. Boone, and S.G. Stafford. 1987. Responses to pathogen-induced disturbance: Decomposition, nutrient availability, and tree vigor. *Forestry* 60:219–227.

Waterman, P.G. 1983. Distribution of secondary metabolites in rain forest plants: Towards an understanding of cause and effect. In: *Tropical Rain Forest: Ecology and Management*, edited by S.L. Sutton, T.C. Whitmore, and A.C. Chadwick, pp. 167–179. Oxford, Blackwell Science.

Whipps, J.M. 2001. Microbial interactions and biocontrol in the rhizosphere. *Journal of Experimental Botany* 52:487–511.

Williams, P.H., S.C. Jarvis, and E. Dixon. 1998. Emission of nitric oxide and nitrous oxide from soil under field and laboratory conditions. *Soil Biology and Biochemistry* 30:1885–1893.

Appendix 2.1

Narratives to Tables 2.2, 2.3, and 2.4

NARRATIVE TO TABLE 2.2 ABOUT GRASSLANDS

- *Food production:* In both unmanaged and managed grasslands, plant products are considered insignificant because plant materials are harvested by herbivores, but they would be affected by biotic and abiotic factors in the same way as animal products. Management shifts decomposition and organic matter transformations toward bacterial dominance, with a reduction of faunal diversity, especially macroarthropods.
- *Water quality:* Water quality refers to runoff to streams. The context is essentially riparian; as in dryland systems, potential evapotranspiration exceeds precipitation. Different dynamics exist in mesic systems. Simplification of soil organisms reduces the retention of nutrients (C, N, and P) in living biomass. Nutrients released or transformed may be directed more into runoff than may percolate to the water table. Higher stocking densities introduce undesirable bacteria into runoff and reduce infiltration by compacting the soil.
- *Water volume:* Factors reducing evapotranspiration will be paramount in dryland systems. These include plant diversity (root depth, root architecture, and hydraulic lift), organic matter stratification and particle size distribution, and biopore formation by macrofauna.
- *Other products:* Fiber production (e.g., hides, carcass contents) is not an objective in managed grasslands, but a by-product. Animal production is roughly inversely proportional to population density in unmanaged grasslands (i.e., it is a function of plant production, which in turn depends on nutrient recycling and transformations by microorganisms).
- *Recreation:* The service rank of unmanaged grasslands for recreation is derived from both wildlife and the aesthetic value of biological diversity and landscape heterogeneity. Managed (and fenced) grasslands are harboring some wildlife, and in some areas access is given to hiking. Access to fenced land may be the subject of legal disputes, and landscape simplification or dissection reduces overall aesthetic value, but this has enormous potential for recreation in many industrialized countries.
- *C sequestration:* A managed grassland means it has been tilled. The C dynamics of untilled pastures are uncertain. Again, a large difference would be expected between the responses of dryland and mesic grassland systems. Accumulation of organic matter at the surface of the soil profile is greater in unmanaged systems with stratification downward and more C directed into complex long-term stable pools by fungal-dominated organisms.
- *Trace gases and atmospheric regulation:* The C and N fluxes of unmanaged grasslands are probably not very significant in terms of the global cycles of these elements, as out-

puts of greenhouse gases to the atmosphere by components of the soil organisms (fungal decomposers, nitrifiers, and denitrifiers) are restricted by corresponding sequestrators (primary producers and nitrogen fixers). Fertilizer input (chemical or animal dung) causes a large increase in both nitrification and denitrification (according to context), from which process greenhouse-forcing NO_x gases are by-products. Disturbance of any kind (including compaction) strongly reduces CH_4 oxidation by archaea in soils.

Narrative to Table 2.3 about Forests

- *Food production:* A side benefit especially of unmanaged forests. Fungal fruiting bodies are a forest food product. Fungal diversity in unmanaged systems may be higher due to the higher diversity of trees and other plants than occurs in managed systems. Many species of soil invertebrates (e.g., ants) are an important food source for birds/wildlife, which in turn are important for recreation. Soil pH is a moderately important determinant of fungal diversity. Managed beech systems also produce truffles.
- *Water quality and volume:* Flood and erosion control. Different plant species have widely different attributes, which can affect soil water quality and quantity (evapotranspiration, different rooting depths/architecture, hydraulic lift). Soil organisms can affect water quality through production of NO_3. Bioturbators affect soil physical properties, which, in conjunction with topography, affect runoff and infiltration.
- *Fuel:* When in the form of wood, fuel is a potentially important service of both unmanaged and managed systems. However, wood fuel may be more commonly extracted from unmanaged systems since managed systems are typically intended for fiber production.
- *Biochemicals and medicines:* Unmanaged forests are used extensively for bio-prospecting, particularly for microbial diversity, genes, and potentially useful products (antibiotics, yeasts, etc.) for industrial or medicinal properties. The diversity of genes, and of useful products, is likely to be related to the diversity of the forest.
- *Habitat provision:* Soil organisms encourage nutrient cycling for plant growth. Ecosystem engineers (e.g., earthworms) create new habitat and provide food for other animals.
- *Waste disposal:* Soil fungi and bacteria affect the accumulation of heavy metals in plants and indirectly into animals. Soil texture and drainage affects a system's ability to hold pollutants, pathogens, and heavy metals. However, we have ranked them with 0, as forests are not intended to be used for waste disposal.
- *Biological control:* Ants predate Lepidoptera pests; mycorrhizae and fungi discourage root pathogens. Water logging of soils encourages fungal pathogens such as root rot.
- *Trace gases and atmospheric regulation:* Soil organisms (nitrifiers, denitrifiers, methane oxidizers) are important to trace gas production and to scrubbing the atmosphere of NO_x, N_2O, SO_2, CH_4, and NH_x. Soil pH, texture, and structure provide anaerobic

microsites for trace gas production. Forest ecosystems are particularly important for methane oxidation. When fertilized or limed, the dynamics of emissions are changed: up to 10 percent of N fertilizers may be denitrified. There exists some doubt as to the organisms responsible for CH_4 in forest ecosystems, but the organisms responsible have been identified as type II methanotrophs. Forest ecosystems are well known to act as sinks for a variety of air pollutants, such as SO_2, NO_x, and NH_x. Diversity has been used as an indicator of aerial ecosystem pollution.

NARRATIVE TO TABLE 2.4 ABOUT ARABLE LAND

- *Food production:* Crop variety, rooting type, and the nature and composition of residues are critical to the quality and quantity of food service provided. Animal production is indirectly affected by the use of arable products for fodder.
- *Water quality:* No-till agriculture is considered to leach fewer nutrients to ground and surface water than occurs when the soil is tilled regularly. Topography is an important factor influencing water runoff, more in tilled than in non-tilled systems: soil tillage leads to exposed soil, which is sensitive to erosion.
- *Water volume:* Non-tillage systems have more biopores formed by earthworms than tilled systems, which benefits water storage volume. Moreover, non-tilled systems are less sensitive to topography than tilled systems because of constant surface cover. On slopes, for example, the direction of soil tillage (along or across altitude lines) is also crucial for runoff of surface water.
- *Other products:* A number of crop plants are used for fiber production (cotton, flax), and effects of biotic and abiotic processes are similar to those for food production.
- *Waste disposal:* Detoxification of waste products is lower under no-till systems because wastes cannot be incorporated, leading to volatilization from the soil surface. The role of soil biota, however, is higher than in tilled systems, where the waste may be directly introduced into the soil. In addition, under no-till systems, concentrations of intermediate toxic products and pesticides can build up in surface soil layers, along with organic matter and nutrients.
- *Biological control:* We define biological control as control of pathogens/weeds by another organism. Rotations in tilled conditions—through maintaining microbial activity and diversity, and through disrupting disease and arthropod cycles and also mycorrhizal networks—improve biological control more in till than in no-till monoculture. However, multi-year rotations may not be economical. Earthworms can have a negative effect on plant parasitic nematodes: for example, in India, joint management of earthworm communities and organic resources doubled tea production while regenerating degraded soils. The effect of earthworms is hypothesized to be obtained through different processes, including suppression of nematode parasites and release of plant growth promoters through enhancement of mycorrhizae. There is little known on the effects of biological agriculture and landscape management (small-scale

or fragmented landscape versus large-scale landscape) on soil-borne disease management. However, whereas no-till conventional agriculture uses herbicides to control weeds, in organic agriculture (e.g., Brazil), cover-crops are used to kill weeds. Effects of GMOs are currently strongly disputed, and the potential solution of GMOs for one problem (weeds) may enhance others (more disease incidence). Therefore, we have weighed the GMO effect neutral.

- *Recreation:* Our concept is habitat for soil biodiversity. The key biotic aspect here is how field margins and riparian areas are managed. Field margins managed for habitat not only harbor diversity, they can also act as refuges for biological control agents, especially predatory arthropods such as beetles. Landscape aspects are important for aesthetic value and crop species matter, since some crops (e.g., corn) do not allow landscape-wide views. Riparian areas managed for habitat enhance surface and groundwater quality. On average, we assessed the recreational value of tilled and non-tilled systems to be equal, especially due to landscape effects.

- *Carbon sequestration:* Carbon sequestration ranks slightly higher in non-tilled than in tilled systems, because the rate of decomposition of crop residues and roots can be slightly less. When the organic matter pool is in balance, effects of C sequestration will be neutral in most cases.

- *Trace gases and atmospheric regulation:* Soil structure is considered under abiotic factors only, though it is clearly a product of both abiotic and biotic factors, especially macro- and microengineers that form aggregates. Specific aspects of macrofauna can alter trace gas emission, for example, denitrification can intensify in earthworm casts.

- *Nutrient cycling:* The microbial community and their activity are essential for nutrient cycling and are moderated by the micro–food web. Synchrony of mobilization and immobilization depends on the dynamics of the micro–food web. The shift from a bacteria-based soil food web under till to a fungal-based food web in no till triggers an associated shift in the nematode and microarthropod assemblage, and alters the micro–food web.

    P-cycling is dependent on soil properties; for example, the amount and nature of clay, the nutrient content of the parent material, and soil enzymes. Cultivation (till) can decrease the enzymes (arylsulfatase and acid phosphatase) involved in S and P transformations. P uptake by mycorrhizae is variable, and is more important in no till systems.

- *Other goods and services:* We have not mentioned habitat provision, biochemicals and medicines, and fuel/energy in the table. These goods and services may indeed be provided by arable land, but these aspects are so context dependent that they are preferably explored in individual case studies.

3

Ecosystem Services Provided by Freshwater Benthos

Alan P. Covich, Katherine C. Ewel, Robert O. Hall, Jr.,
Paul S. Giller, Willem Goedkoop, and David M. Merritt

The concept of ecosystem goods and services (Daily 1997; Heal 2000; Brismar 2002) conveys how natural processes such as biomass production and nutrient cycling are essential to the Earth's capacity for sustaining human populations. Here we examine how species diversity and ecosystem processes, which supply these goods and services to human societies, are mediated by sediment- or bottom-dwelling (benthic) organisms in fresh waters. Benthic invertebrates, microbes, and aquatic plants are widely distributed in fresh waters. Their ecology is well understood in many temperate-zone regions and the diversity of freshwater benthic communities is broadly documented (Bronmark & Hansson 1998; Giller & Malmqvist 1998; Thorp & Covich 2001). This biota includes some species that are widespread, functional generalists and others that are restricted in their distributions and are functionally specialized.

Sediment-dwelling plants and invertebrates provide numerous critical ecosystem services in fresh waters (Ewel 1997; Covich et al. 1999), yet economic valuation of associated ecosystem functions is rarely measured other than in shellfisheries production (Carpenter & Turner 2000; Odum & Odum 2000). What are the values of nonmarket goods and services derived from a lake or river or wetland? Relationships between species diversity, water resource allocations, and freshwater ecosystem services are being evaluated by ecologists and economists (Loomis 2000; Daily & Ellison 2002; National Research Council in press). Moreover, benthic biologists are beginning to determine how the loss of different species affects freshwater ecosystem functioning (Wall et al. 2001); to date, however, these experiments have primarily focused on small-scale, short-term studies of relatively few species (Giller et al. 2004; Covich et al. in press). Vulnerability of ecosystem services is increasing because of the elimination of many fresh-

The authors would like to acknowledge Nina Caraco for her contributions to this chapter.

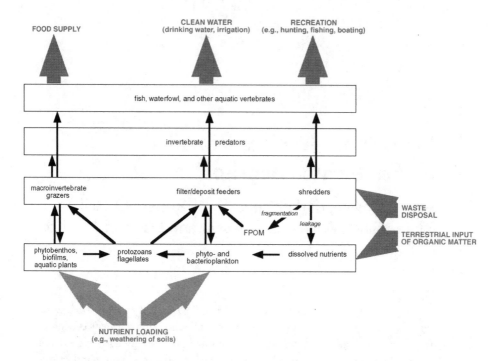

Figure 3.1. Schematic overview of a functional food web showing the linkages between food web processes and services provided by sediment and above-surface biota in freshwater ecosystems. Small black arrows indicate pathways for nutrient uptake and cycling from sediment-dwelling species to those in the open waters. Large gray arrows indicate major nutrient sources (inputs) and resulting ecosystem services (outputs) from biological processing in lakes and streams. FPOM = Fine Particulate Organic Matter.

water habitats and the accelerated rates of extinction among key species. Time for research on the ecological and economic importance of species is short (Everard & Powell 2002; Dudgeon 2003), and there is growing concern about losing these services following declines in species diversity (Davies & Day 1998; Moulton 1999).

In this chapter we outline the types and importance of freshwater ecosystem services. In particular, we discuss the role of benthic species in ecosystem processes such as productivity, nutrient transformations, and decomposition of organic matter (Figure 3.1). We also examine patterns of inter-connected ecosystem processes and ways to evaluate them. We briefly review how particular sediment-dwelling organisms can alter freshwater ecosystem services. We discuss three major categories of ecosystem services: provisioning, supporting, and cultural (Millennium Ecosystem Assessment 2003). Each of these categories is then illustrated with examples of how benthic species provide different ecosystem services. The vulnerability of ecosystem services in fresh waters is discussed in Chapter 6, where we consider the concept of disservice (or the exploitation of one ecosystem service that leads to a negative effect or elimination of a second service) and review several case studies.

Importance of Freshwater Ecosystem Services and Biodiversity

Ecosystem services in fresh waters depend on a range of different benthic species (Tables 3.1a–3.1e). For example, fish and shellfish yields depend heavily on sustained production of diverse benthic prey species (Huner 1995; New & Valenti 2000). Although only a few of the 390 species of crayfish native to North America are harvested for food, other crayfish species play major roles in ecosystem dynamics by linking sedimentary habitats with overlying waters through burrowing and mixing of sediments, nutrient cycling, breaking down dead organic matter, and grazing on submerged macrophytes (Covich et al. 1999; Hobbs 2001). Benthic invertebrates are essential prey for bottom-feeding fishes and aquatic mammals such as river otters and raccoons. Other freshwater ecosystem services include the breakdown of industrial and residential wastes by microbes and invertebrates (Geber & Bjorklund 2001; DeBruyn & Rasmussen 2002). Fresh waters dilute wastes, provide cooling waters for power generation and other industrial processes, as well as serve demands for recreational swimming, fishing, and boating (Postel & Carpenter 1997). The economic values of fisheries (New & Valenti 2000; Welcomme 2001) and recreational activities in fresh waters are well documented (Loomis 2000). Moreover, managers rely on monitoring services provided by benthic invertebrates by measuring changes in benthic species' presence and abundance to quantify indicators of water quality (Johnson et al. 1992; Clements & Newman 2002). These benthic species integrate local impacts over various time scales and provide important information on concentrations of dissolved oxygen, nutrients, and specific toxins. All these services are important for managerial decisions regarding water allocations (Postel & Carpenter 1997; Strange et al. 1999).

Benthic ecosystem services are sometimes considered a free resource (Mitsch & Gosselink 1993; Barbier et al. 1994; Acharya 2000). For example, clean drinking water supplies are derived from natural watersheds (Watson & Lawrence 2003). This essential ecosystem service is well studied by ecologists, economists, and environmental engineers. Clean drinking water can be naturally sustainable because of the role played by benthic species that carry out biofiltration, detoxification, and numerous processes that break down organic wastes in rivers, lakes, and groundwaters. Without this recycling by diverse microbes and benthic invertebrates, organic matter accumulates and leads to deoxygenation (through microbial respiration), which then causes rapid deterioration in water quality and often results in fish kills. The effectiveness of this natural "self-cleaning" ecosystem service is limited by the quantity, type, and rates of organic waste inputs that can be processed biologically under specific flow conditions and retention times (see Giller et al., Chapter 6). Thus, anthropogenic threats and influences

Continued on page 54

Table 3.1a. Positive ecosystem service rankings (relative within fresh waters; not quantitative) for managed (flood-control reservoirs) and unmanaged lakes.

Explanation of service values:
(−3) = strong disservice; (0) = neutral; (3) = strong positive; n.a. = not applicable.

Goods and Services	Lakes: Unmanaged			Lakes: Managed		
	Service Rank	Biotic	Abiotic	Service Rank	Biotic	Abiotic
Provisional Services						
Food production						
Plant	0	n.a.	n.a.	0	n.a.	n.a.
Animal	2–3	***	*** flow-dependent	0–1	***	*** naturally flow-dependent
Other products						
Fuel/Energy	1	n.a.	***	1	n.a.	***
Fiber	0	n.a.	n.a.	0	n.a.	n.a.
Potable water	3	***	** dilution, sorption	0	n.a.	n.a.
Water quantity	3	n.a.	***	3	n.a.	***
Supporting Services						
Waste processing	2	***	*** dilution temperature	1	n.a.	n.a.
Climate modification C sequestration	0	n.a.	n.a.	0	n.a.	n.a.
Trace gas production	−1	***	* temperature	−1	***	*
Irrigation	3	n.a.	***	3	n.a.	***
Transport	3	*	***	3	*	***
Cultural Services						
Recreation	3	***	***	1 fish/boat/fowl	*	3
Aesthetic						

Asterisks indicate the relative importance of biotic and abiotic factors, from weak (*) to strong (***), in the provision of the associated good or service.

Table 3.1b. Positive ecosystem service rankings (relative within fresh waters; not quantitative) for rivers and waterways.

Explanation of service values:
(−3) = strong disservice; (0) = neutral; (3) = strong positive; n.a. = not applicable.

	Rivers: Unmanaged			Waterways: Managed (Channelized)		
Goods and Services	Service Rank	Biotic	Abiotic	Service Rank	Biotic	Abiotic
Provisional Services						
Food production						
Plant	0	n.a.	n.a.	0	n.a.	n.a.
Animal fish shellfish	2–3	*** decomposers, organic matter transformation, nutrient transformation (nitrification, etc.), intact food web	** trophic status, nutrient inputs, temperature	0–1	*** modified food web	*** chemical inputs, depth, geomorphology, topography, precipitation, trophic status, nutrient inputs, temperature
Other products						
Fuel/Energy	0	n.a.	n.a.	0	n.a.	n.a.
Fiber	0	n.a.		0	n.a.	n.a.
Potable water	3	*** nutrient cycling, decomposition, nitrification, phosphate liberation, DON, DOC	** residence time, depth topography	3	*** nitrification, phosphate liberation, DON, DOC	**

(continued)

Table 3.1b. *(continued)*

Goods and Services	Rivers: Unmanaged			Waterways: Managed (Channelized)		
	Service Rank	Biotic	Abiotic	Service Rank	Biotic	Abiotic
Water quantity	2–3	*	***	3	organic matter, moisture retention *	geomorphology *
Supporting Services						
Waste disposal	3	nutrient cycling, decomposition ***	**	1–2	***	**
Climate modification	2	organic matter burial in sediments, decomposition ***	Calcium carbonate deposition **	1	bacterial-based food webs enhance C loss (but still net C accumulation), CH_4 emission from manure **	moisture, complexing, etc. *
C sequestration	1		**	1		
Trace gas production	–1					
Irrigation	1	n.a.	***	3	n.a.	***
Transport	3	***	**	1–2	*	***
Cultural Services						
Recreation	3	***	***	1	fish/boat/fowl *	3
Aesthetic						

Asterisks indicate the relative importance of biotic and abiotic factors, from weak (*) to strong (***), in the provision of the associated good or service.

Table 3.1c. Positive ecosystem service rankings (relative within fresh waters; not quantitative) for managed and unmanaged wetlands.

Explanation of service values:
(−3) = strong disservice; (0) = neutral; (3) = strong positive; n.a. = not applicable.

Goods and Services	Wetlands: Unmanaged			Wetlands: Managed		
	Service Rank	Biotic	Abiotic	Service Rank	Biotic	Abiotic
Provisional Services						
Food production						
Plant	1	**	***	3	**	***
Animal	2	***	***	1	***	***
Other products						
Fuel/energy	0	n.a.	n.a.	0	n.a.	n.a.
Fiber	2	**	***	0	**	***
Potable water	3	***	***	0	n.a.	n.a.
Water quantity	0	n.a.	n.a.	−1	n.a.	***
Supporting Services						
Waste disposal	3	***	***	2	***	** application of manure
Climate modification						
C sequestration	0	n.a.	n.a.	0	n.a.	n.a.
Trace gas production	3	***	***	3	***	***
Irrigation	0	n.a.	n.a.	0	n.a.	n.a.
Transport	0	n.a.	n.a.	0	n.a.	n.a.
Cultural services						
Recreation	1	***	***	0	n.a.	n.a.
Aesthetic	2	**	***	2	***	***

Asterisks indicate the relative importance of biotic and abiotic factors, from weak (*) to strong (***), in the provision of the associated good or service.

Table 3.1d. Positive ecosystem service rankings (relative within fresh waters; not quantitative) for groundwater and abstraction.

Explanation of service values:
(−3) = strong disservice; (0) = neutral; (3) = strong positive; n.a. = not applicable.

Goods and Services	Groundwater: Unmanaged			Abstracion: (Artificially Recharged)		
	Service Rank	Biotic	Abiotic	Service Rank	Biotic	Abiotic
Provisional Services						
Food production						
Plant	0	n.a.	n.a.	0	n.a.	n.a.
Animal	0	n.a.	n.a.	0	n.a.	n.a.
Other products						
Fuel/energy	1		*** geothermal	1		*** geothermal
Fiber	0	n.a.	n.a.	0	n.a.	n.a.
Potable water	3	***	** percolation	3	***	*** infiltration (recharge)
Water quantity	3	n.a.	***	3	n.a.	***
Supporting Services						
Waste disposal	1	** biore-mediation	** reverse wells	0	n.a.	n.a.
Climate modification	0	n.a.	n.a.	0	n.a.	n.a.
C sequestration	0	n.a.	n.a.	0	n.a.	n.a.
Trace gas production	0	n.a.	n.a.	0	n.a.	n.a.
Irrigation	3	n.a.	***	3	n.a.	***
Transport	0	n.a.	n.a.	0	n.a.	n.a.
Cultural Services						
Recreation	0	n.a.	n.a.	0	n.a.	n.a.
Aesthetic	1	* microbial mats, sulphur precipitation	geysers, springs	0	n.a.	n.a.

Asterisks indicate the relative importance of biotic and abiotic factors, from weak (*) to strong (***), in the provision of the associated good or service.

Table 3.1e. Positive ecosystem service rankings (relative within fresh waters; not quantitative) for prairie floodplain and drained wetlands.

The column on the left represents an intact ecosystem and the one on the right represents a managed ecosystem likely to be derived from the other. Explanation of service values: (−3) = strong disservice; (0) = neutral; (3) = strong positive; n.a. = not applicable.

Goods and Services	Prairie Wetlands and Floodplain Forests			Agricultural Crops on Drained Wetlands		
	Service Rank	Biotic	Abiotic	Service Rank	Biotic	Abiotic
Provisional Services						
Food production						
Plant	0	n.a.	n.a.	3	**	***
Animal	2	***	**	2	***	**
Other products						
Fuel/energy	1	***	**	2	***	**
Fiber	0	n.a., no present-day use	n.a.	0	n.a.	n.a.
Potable water	3	**	***	−2	n.a.	***
Water quantity groundwater recharge flood mediation	2	n.a.	***	−1	n.a.	***
Supporting Services						
Waste disposal Climate modification	0	n.a.	n.a.	0	n.a.	n.a.
C sequestration	0	n.a.	n.a.	0	n.a.	n.a.
Trace gas production	1	***	***	0	n.a.	n.a.
Irrigation	0	n.a.	n.a.	0	n.a.	n.a.
Transport	0	n.a.	n.a.	0	n.a.	n.a.
Cultural Services						
Recreation	2	***	**	2	***	**
Aesthetic	3	***	**	1	**	**

Asterisks indicate the relative importance of biotic and abiotic factors, from weak (*) to strong (***), in the provision of the associated good or service.

can alter the balance of natural regulatory factors such as energy flow, organic matter transport, hydrologic regimes, biogeochemical cycles, and hydrochemistry (Palmer et al. 2000; Malmqvist & Rundle 2002). They change the structure of sediments, alter temperature regimes, and cause other extreme environmental conditions beyond normal levels of variation.

Misuse or overuse of one type of ecosystem service can lead to a negative effect on other important services and the biota and the ecosystem functions that underpin them (see Giller et al., Chapter 6). For example, soils are critical in the production of food and fiber, but overexploitation of terrestrial ecosystem services can diminish downstream services provided by sediment-dwelling systems. Runoff from intensive agricultural fields following heavy rains can contain excessive nitrogen because of the overuse of fertilizers or the accumulation of animal wastes. When these high-nutrient concentrations are combined downstream with nitrogen and phosphorus from sewage effluents and other sources, they cause excessive growth of aquatic plants and deoxygenation associated with eutrophication. Human health suffers from the poor water quality resulting from growth of toxic algal species (Burkholder 1998; Anderson et al. 2002) and increased abundance of disease pathogens in nutrient-laden rivers and estuaries. Furthermore, deoxygenation of nutrient-rich fresh waters results in increased ammonia, which is toxic to fish and many benthic invertebrates. The buildup of nitrate in groundwater can also pollute drinking waters and result in "blue babies" (methaemoglobinaemia) when infants drink contaminated water (Bouchard et al. 1992; Gupta et al. 2000; Mallin 2000). Thus, the provision of clean drinking water (through biotic treatment in ground waters and surface waters) is lost because of eutrophication (Brock 1985; Baerenklau et al. 1999; Boyle et al. 1999; Carpenter et al. 1999; Bockstael et al. 2000). Disservice results when ecosystems are poorly managed and positive natural processes are lost (see Giller et al., Chapter 6). Protection of catchments and good riparian and wetland management contributes to the maintenance of ecological processes and associated critical ecosystem services.

Analysis of Roles Played by Benthic Species

1. *Ecological evaluation of ecosystem service.* Society has alternative uses for fresh water, which are associated with competing demands for particular quantities and qualities of water. For example, in Central Asia, increased diversions of water from the Amu and Syr Darya Rivers expanded production of irrigated agriculture and other upstream uses beginning in the early 1960s, but resulted in major declines in fish production and water-based transportation after the Aral Sea partially dried and became more saline. The results included loss of aquatic species and endangerment of human health in and around what had been the fourth largest lake in the world (Williams 2002). Another example of these trade-offs for competing demands for fresh waters led to ecosystem degradation resulting from lower lake levels in Mono

Lake, east of the Sierra Nevada Mountains of California. Demands for fresh water increased rapidly in the Los Angeles Basin and water was diverted from the Owens River in the Mono Lake Basin. These diversions to Los Angeles resulted in fewer breeding sites for migratory waterfowl and changes in lake food webs as salinity increased (Hart 1996). Local community action eventually restored the integrity of the lake ecosystem (Loomis 1987, 1995). These sorts of trade-offs require careful ecological evaluation of the full range of ways in which water allocations and species loss may alter ecosystem services. Ecologists and economists are beginning to quantify trade-offs among different uses of ecosystems influencing water quality and yield (Whigham 1997; Loomis et al. 2000). As is discussed below, more research is needed to evaluate how loss of species diminishes or eliminates critical ecosystem services.

2. *Economic evaluation of ecosystem services.* Many natural freshwater ecosystem processes have definable economic values (Abramovitz 1998; Pearce 1998) as well as the non-use, existence, and aesthetic values that must also be evaluated in ways that reflect the importance of protecting benthic species and their habitats. Methods to determine economic values include market pricing, contingent evaluation, cost-benefit analysis, consideration of replacement costs, and prices of substitute goods and services, if any (National Research Council in press). Values of ecosystem services such as the production of high-quality drinking water or storm mitigation (protection of river banks and lake shores by riparian vegetation) can be estimated by determining how much people are willing to pay for, or, if possible, to replace these services (Cleveland et al. 2001; Daily & Ellison 2002). The costs of comparable substitutes (e.g., in the form of engineered replacements for natural benthic ecosystems) provide one means to evaluate the economic values of some natural services provided by sediment-dwelling organisms. For example, building filtration plants to provide clean drinking water for New York City could cost from US$2 to 8 billion (Foran et al. 2000; O'Melia et al. 2000; Gandy 2002), while protection of the watershed's natural communities of benthic invertebrates and improved riparian management is likely to save many or all of the costs of building filtration plants (Daily & Ellison 2002). On-going studies of water quality and stream invertebrates by ecologists at the Stroud Water Research Center in Pennsylvania, USA, are documenting these ecosystem services (Bern Sweeney, personal communication 2003). Replacement costs of wetlands that provide natural filtration are also generally large (Mitsch & Gosselink 1993; Bedford 1996; Williams 1999) because of the complex nature of these ecosystem processes performed by numerous sediment-dwelling species of plants and animals. For example, more than US$500 million is being spent to restore 11,500 hectares along 90 km of the old river channel of the Kissimmee River Basin in Florida (Dahm et al. 1995; Toth et al. 1998). Services from the natural meandering river and its floodplain were lost in the 1960s when 167 km of the river was channelized and 21,000 hectares of wet-

lands were drained (Whalen et al. 2002). Other attempts to replace lost services with artificially constructed wetlands have had limited success, especially if native species and habitat structures are not included in the design (Zedler 2000; Bonilla-Warford & Zedler 2002; Stevenson & Hauer 2002). Similarly, as discussed in Chapter 6, attempts to restore ecosystem services provided by benthic organisms in European rivers (following industrial pollution, channelization, and dam building) continue to face serious constraints (Cioc 2002). In many cases there are no satisfactory and sustainable substitutes for natural ecosystem services.

Complexity of Natural and Managed Freshwater Ecosystems

Determining the total economic value of freshwater ecosystems is very difficult because some of their ecological functions have competing commercial values and others have primarily aesthetic or existence values. Studies that illustrate the importance of freshwater benthic ecosystems in providing essential services for sustainable human populations require a comprehensive perspective on evaluation of freshwater services that include both use and non-use values (National Research Council in press). The values of drinking water, freshwater fish, and shellfish, as well as recreational uses of rivers, lakes, and wetlands are very important to humans regardless of the methods used to estimate their market values. Indeed, "use values" derived from market pricing and the intrinsic "existence values" (estimated from surveys and nondirect methods) can be complementary and combined to document the need to maintain the biodiversity of fresh waters. Once species are lost, their economic values become abundantly clear to the general public but their natural services often cannot be fully restored or artificially replaced. Lessons from these failures can be extended to avoid similar future losses in other freshwater ecosystems.

Types of Freshwater Ecosystem Services

The goods and services provided to humans by freshwater benthic ecosystems may be classed as *provisioning services*, or products obtained from ecosystems, such as plant and animal food and fiber; *supporting services*, or services necessary for the production of all other ecosystem services, such as waste processing, the production of a sustained clean water supply, flood abatement, and climate moderation; and *cultural services*, or nonmaterial benefits obtained from ecosystems, such as aesthetics, education, and recreation (Millennium Ecosystem Assessment 2003). Besides natural waste treatment that enhances water quality, many freshwater ecosystems are critical habitats for certain life stages of marine and freshwater fishes, waterfowl, and other sources of human foods. Moreover, these benthic ecosystems provide critical habitat for many other species.

Understanding natural processes that contribute to ecosystem services is of immediate concern given the rates at which human activities are altering natural fresh waters.

As human population densities increase further and new chemical compounds and technologies are developed, unanticipated consequences will have long-lasting impacts on freshwater benthic ecosystems (Malmqvist & Rundle 2002). If fresh waters are degraded under intensive exploitation, their natural processes can be diminished or lost completely. As native species are lost through local extinction and nonnative species are introduced into fresh waters, there is lively debate regarding trade-offs among different management alternatives. It is critical that decision makers understand how species can provide unique roles in cycling nutrients and in producing valuable commodities and services in many different types of fresh waters. Tables 3.1a–3.1e highlight the major goods and services under a number of major categories (including food production, water quality and quantity, waste disposal, climate modification, and recreation). The relative importance of services varies across the different freshwater ecosystems.

Groundwater

Provisioning Services. Groundwater supplies drinking, municipal, industrial, and irrigation water worldwide. The most important ecosystem service humans receive from groundwater is providing clean water for drinking. Although abiotic processes control water quantity through recharge, microbes are especially important in producing clean water. Microbial remediation of contaminated groundwater is another ecosystem service provided by sediment-dwelling organisms. Bioremediation of groundwater often benefits from injecting microbial assemblages into contaminated sites and encouraging bacterial growth with nutrient additions (Ghiorse & Wilson 1988; Baker & Herson 1994). For example, bacteria can remove nitrate or degrade recalcitrant organic contaminants. Groundwater supports a rich food web consisting of microbes and metazoan consumers (Marmonier et al. 1993). Because these groundwater species respond to chemical contamination, they can be used to identify polluted aquifers (Gounot 1994; Moeszlacher 2000). For example, the presence of certain flagellates and their grazing of bacteria may increase degradation rates of toluene (Mattison & Harayama 2001). However, little is known about the details of food web interactions in groundwater or subsurface flows in stream channels (hyporheic zones) that alter degradation and removal rates of contaminants (e.g., nitrate, organic compounds).

Lakes and Rivers

Provisioning Services. Benthic organisms in lakes and rivers provide food production mostly through the dependence of fish production on invertebrate prey and nutrient cycling. Globally about 8×10^6 tons of freshwater fish are harvested, with double that amount produced by aquaculture (FAO 1995). Productivity of these fisheries will, in part, depend upon benthic production directly (e.g., consumption of benthic invertebrates or aquatic plants) or indirectly (e.g., benthic mineralization of nutrients). For example, Chinese polyculture relies on benthic productivity for plant and mussel production for feeding carp. Productivity of deepwater ecosystems is often influenced by how tightly the upper waters are linked to nutrient cycling in the lower waters that are in contact with sedimentary sources of nutrients. This process of pelagic-benthic coupling is critical in determining how nutrients (or toxins) are stored in sediments and seasonally cycled into surface waters, where they are incorporated into algal production and then consumed by filter-feeding zooplankton and fishes. Several marine fisheries influence and depend upon production in freshwater ecosystems (e.g., anadromous salmon spawn and juveniles are reared in freshwater rivers and lakes).

Support Services. Benthic species maintain water quality via transformation of excess nutrients and organic pollutants. For example, stream organisms can rapidly take up and incorporate nitrogen into their biomass or produce ammonia or methane that enters the atmosphere, thereby lowering the loads of dissolved organic nitrogen (Alexander et al. 2000). Nutrients such as phosphorus in streams can be temporarily stored in sediments and the biota (Meyer & Likens 1979). Benthic bacteria permanently remove nitrogen via denitrification (e.g., Pind et al. 1997) as well as convert nitrogen from unusable to usable forms that can be taken up during plant growth. Invertebrates and microbes that are widely distributed in natural ecosystems also occur in sediments and biofilms found in water-treatment plants. Thus, the biological functions are similar, although the species and densities generally vary greatly between natural and artificial habitats and these communities respond primarily to nutrient loading (e.g., Kadlec & Knight 1996; DeBruyn & Rasmussen 2002). Other species of macroinvertebrates (such as stoneflies and caddisflies) that break down particular types of organic materials are restricted in their distributions. Many occur in pristine, unmanaged habitats where low levels of nutrients and high concentrations of dissolved oxygen are sustained by a diverse assemblage of plants and animals.

The breakdown of dead organic matter (detritus) is an ecosystem service provided by most freshwater benthic communities. The roles of benthic detritivores that transform and transfer nutrients are well documented (Wallace et al. 1996; Giller & Malmqvist 1998). Benthic organisms shred coarse sizes of organic matter into finer particles. Microbial species condition detritus, which facilitates use by the shredding invertebrates, and also decompose organic particles. Microbes also produce gases (CO_2, CH_4, N_2) that enter the atmosphere and dissolved forms of nutrients that enter the overlying waters. Dissolved nutrients increase the growth of algae and aquatic macrophytes, which in turn are consumed by herbivorous and omnivorous invertebrates and fishes, thus creating the basis for complex food webs (Covich et al. 1999; Crowl et al. 2001; Jonsson & Malmqvist 2003).

As previously discussed, water quality is maintained by a number of biotic processes that are associated with sedimentary habitats where benthic invertebrates play well-defined roles in ecosystem processes. The loss of certain species or their changes in abundance may impair ecosystem function and consequently ecosystem services. For example, findings from Coweeta Hydrologic Laboratory, a US National Science Foundation Long Term Ecological Research site in North Carolina, indicated that measures of stream water quality (associated with rates of detrital processing) declined when stream insects were experimentally eliminated from a stream. Sequential declines in aquatic insect biodiversity correlated with the changes in stream ecosystem processes. This study was the first field experiment to show that these measures of water quality correlated with ecosystem processes (Wallace et al. 1996). This research also showed specifically how physical and chemical impacts (which deplete invertebrate populations) may feed back to alter stream ecosystem processes. The ecosystem-scale evidence for this linkage in streams and rivers obtained from the research at Coweeta provided detailed information about specific ways in which ecosystem-level processes change following invertebrate removal. In these studies, many species of leaf-shredding invertebrates were known to process coarse leaf-litter inputs from riparian zones into smaller particles. To test the importance of this role of shredders in ecosystem function, the Coweeta researchers experimentally removed most stream-dwelling insects using low doses of insecticide, which lowered shredder secondary production to 25 percent of that of a nearby reference stream (Lugthart & Wallace 1992). Organic carbon export from the watershed decreased dramatically following the insecticide treatment (Cuffney et al. 1990). Leaf decomposition was twice as slow in the invertebrate removal stream, so standing stocks of leaf litter were much higher. In general, lower export of organic carbon from headwater streams may lower animal production in downstream food webs, where filter-feeding species may be facilitated by shredding species living upstream (Heard & Richardson 1995).

Studies at the Luquillo Experimental Forest in Puerto Rico (another US National Science Foundation Long Term Ecological Research site) further demonstrate the potential for a single species to have an impact on ecological functions. A freshwater shrimp (*Xiphocaris elongata*) is one of the few species of shredders that facilitates the uptake of suspended organic particulates by a filter-feeding species of shrimp (*Atya lanipes*), which co-occurs in some tropical headwater streams (Crowl et al. 2001). Loss of these species' functions of shredding and filter feeding would likely result in slower rates of leaf litter breakdown and less energy flow in the headwater food web. This loss of species that shred leaf detritus may be critical in tropical headwaters, where species of shredding insects are relatively rare and the functional redundancy among leaf shredders is relatively low (Covich et al. 1999; Dobson et al. 2002). Related research is beginning to identify the degree to which ecosystem services of rivers and other freshwater ecosystems are altered directly by physical and chemical impacts (e.g., low O_2, low pH, or high sedimentation) compared with being altered indirectly through the loss of key animal taxa (Jonsson & Malmqvist 2003).

Accumulation of organic matter can slow decomposition by microbial species and detritivorous invertebrates when dissolved oxygen is depleted by high rates of respiration, especially at warm temperatures. Deoxygenation subsequently results in displacement of numerous species that require high oxygen concentrations and replacement by other species that can tolerate the stressful conditions of low dissolved oxygen. This sequence of species substitutions typically results in a degraded stream community with nuisance and disease-transmitting characteristics as well as reduced capacity for providing critical ecosystem functions. For example, heavy pollution and deoxygenation in some urbanized streams around Rio de Janeiro eliminated Atyid shrimp, which previously filtered out suspended organic matter. Following this pollution and loss of freshwater shrimp in the streams, the increased densities of filter-feeding blackflies led to more biting insects due to the loss of the same ecological function (filtration of suspended organic matter) by the shrimp (Moulton 1999). Other benthic invertebrates directly serve as biocontrol agents by feeding upon vectors of diseases (e.g., aquatic insects and crustaceans that feed on certain species of mosquito larvae and snails) that are prevalent in tropical freshwater habitats. Field studies substantiate the widespread importance of benthic invertebrates as indicators of water quality and as functional regulators of important ecosystem functioning (Clements & Newman 2002).

Wetlands and Associated Freshwater Habitats

Among the most critical and scarce freshwater ecosystems are marshes, floodplains, and swamps. Although they cover only roughly 6 percent of the Earth's land surface and are most common in temperate and boreal regions, wetlands perform a wide range of ecosystem functions, many of consequence on a global scale. Most of these functions are related either directly or indirectly to the activities of the flora and fauna living in sediments. Wetlands occur where saturation or inundation often produces anaerobic sediments, limiting rooted plant diversity to only those species adapted to anoxic conditions (Ewel 1997; Brinson & Malvarez 2002). Seasonal and interannual patterns of hydrologic regime and water source (rain, groundwater, and/or riverine surface water) govern many of the characteristics of wetland ecosystems, including species diversity and primary productivity. Freshwater wetland types include wet meadows, fens, bogs, lake margins, floodplain forests and bottomland swamps, tropical peat swamps, and extensive boreal peatlands. All wetlands are flooded long enough to influence the types of biota able to inhabit the site and the character and rate of biogeochemical processes. The global diversity of wetlands derives from regional and local differences in hydrologic regime (especially duration of flooding but also water residence time and water chemistry), physical factors such as fire and storms, unique characteristics of the plant species inhabiting those wetlands, and the influence of the animals that visit and live in them.

These many types of wetlands often are connected to surface and subsurface waters. Their ecosystem services include cycling of nutrients, breakdown of organic matter, and filtering of sediments that otherwise would enter rivers (Naiman & Decamps 1997; Keddy 2000). Yet, these critical habitats are being lost at a rapid rate despite their recognized values and legal standing (Dahl et al. 1991; Bedford 1999; Brinson & Malvarez

2002). Threats to rivers, floodplains, and lakes are also increasing (see Giller et al., Chapter 6) and are likely to result in loss of their essential ecosystem services.

The draining of wetlands and other threats to freshwater ecosystems have given rise to local and regional programs aimed at reducing their loss and restoring them to natural levels of diversity (Zedler 2000; Mayer & Galatowitsch 2001). In some cases new wetlands are constructed in other areas to attempt to offset the loss of natural wetlands (Moshiri 1993; Kladlec & Knight 1996). These constructed wetlands for "mitigation banking" can provide some ecosystem services, but often lack the biodiversity as well as the hydrologic regime that characterize natural ecosystems. Successful management for the sustainability and reliability of ecosystem services remains uncertain.

Provisioning Service. Riparian wetlands often have higher concentrations of microorganisms, insects, and animals than adjacent ecosystems (Naiman & Decamps 1997), and in arid regions they may be the only forested natural vegetation, thereby providing valuable habitat for arboreal species (National Research Council 2002). Many terrestrial animals, both vertebrates and invertebrates, use wetlands during some portion of their lives, and 50 percent of the 800 species of protected migratory birds in America rely on wetlands for habitat and food resources associated with benthic production of invertebrates and aquatic plants (Wharton et al. 1982). For example, 50 to 80 percent of the duck populations in North America are produced in north-central prairie potholes. These ecosystems provide hunters with significant recreational opportunities of economic importance (Batt et al. 1989). Waterbirds use a range of habitats including ponds, swamps, lagoons, mudflats, estuaries, embayments, and open shores of lakes, rivers, and reservoirs. Wetlands flooded to average depths of 15 to 20 cm (fringe and depressional wetlands) accommodate the greatest richness and abundance of birds (Taft et al. 2002).

Beavers play an important role in wetland landscapes as ecosystem engineers, creating a tremendous expansion of wetlands that otherwise would not have existed. Beaver harvests have averaged 400,000 pelts per year over the past century in North America (Novak et al. 1987). Alligators are also harvested for their pelts and meat, generating over US$16 million in a single year in the state of Louisiana, USA (Mitsch & Gosselink 2000). Crayfish aquaculture has also become an important use of natural and created shallow marshes in North America, northern Europe, and Australia in recent years. *Procambarus clarkii* accounts for about 90 percent of the 60–70,000 tons of crayfish cultured annually for food in North America (Huner 1995).

Nearly all commercially harvested freshwater fish and shellfish species depend on fringe or riverine wetlands at some life stage (typically for spawning or for nursery habitat). Anadromous fishes are less reliant directly on freshwater marshes, but fry may use riverine marshes for protection. Plant foods are harvested from fringe, riverine, and depressional wetlands as well as from extensive peatlands. For instance, berries from boreal peatlands are an important and nutritious part of the diet typical of high-latitude human populations (Usui et al. 1994). The worldwide average annual harvest of blueberries (*Vaccinium myrtillus*) was 157,128.6 million tons (1990–2002), with approximately 42,000 ha in production (FAO 2003). The total wild berry harvest in Finland can be as high as 10^9 kg per season for a market value of more than US$240,000 (Wallenius 1999). Rice production in managed wetlands plays an important role in world nutrition and in the global economy. About 596 million tons of rice are produced each year (86 percent of this is consumed by human populations), harvested from 1.6 million km^2 of wetlands (IRRI 2000).

Wetland timber is harvested for pulp and building materials; peat (partially decomposed organic material) for fuel and horticultural soil amendment; and herbaceous vegetation from marshes for livestock fodder, fuel, fiber, and other products. Harvesting may require lowering of water tables to facilitate access to and removal of materials, which may permanently alter species composition. Peat harvesting is often viewed as renewable, but recovery may take centuries or more. Peatlands cover 420 million ha globally, with the most extensive habitats located in Russia and Canada. Peat is used as fuel to generate electricity or for conversion to methanol or industrial fuels (Rydin et al. 1999; Mitsch & Gosselink 2000). It may also be used to remove toxic materials and pathogens from wastewater and sewage (Jasinski 1999). Wetland meadows of many kinds are used for harvesting fodder and grazing livestock throughout the world. In Scandinavia, wet meadows bordering lakes and rivers are some of the most productive areas for the production of livestock fodder (Nilsson 1999; Rosén & Borgegård 1999).

Supporting Service. Wetlands can recharge local and regional shallow groundwater water systems; small wetlands can be very important locally (Weller 1981). Many wetlands may also improve water quality by removing organic and inorganic materials from inflowing waters. Wetland vegetation takes up and stores nutrients and some toxic compounds, thereby removing them from rapid cycling. Where water levels fluctuate, microbial denitrification can reduce nitrogen loads.

Waste processing is a service most often attributed to wetlands, although it is generally restricted to a few kinds of wetlands that can treat only certain wastes under specific conditions. Generally, riverine and fringe wetlands treat non-point-source pollution, such as from agricultural fields, either directly through the uptake of nutrients, chemicals, and metals, or indirectly through the chemical transformation and processing of toxic compounds. For example, the freshwater tidal marshes of the Hudson River retain nutrients and result in denitrification when properly managed (Zelenke 1998). Depressional wetlands and extensive peatlands can substitute for tertiary wastewater treatment (e.g., Odum 1984; Ewel 1997), but the lack of control over waste processing has made construction of artificial wetlands more attractive (Ewel 1997). Wetlands created for further treatment of secondary sewage from major cities can remove up to 97 percent of the nitrogen delivered to them through a combination of uptake by plants and through denitrification (Costa-Pierce 1998).

The ability of wetlands to process wastes effectively depends on the rates of nitrogen, iron, manganese, sulfur, and carbon transformations that occur under increasingly low oxygen conditions in the sediments. Although wetlands maintain the widest range of oxidation-reduction reactions of any ecosystem, effective waste processing depends on appropriate ratios of many compounds. Overloading the system can compromise ecosystem functions. Waste processing and biological fixation of nitrogen relies on microbes such as *Azotobacter, Clostridium butyricum, Rhizobium* in root nodules, and cyanobacteria. Sediment-dwelling fauna affect surface and subsurface flows of water as well as stimulate microbial activity, even to the extent of changing the entire nature of a wetland. Beavers dam rivers, creating ponds and fringe wetlands, and alligators excavate cavities in wetlands in karst regions, such as the Florida Everglades (United States), facilitating the concentration of fish in patches of swamp wetlands during dry seasons.

Large expanses of wetlands (extensive peatlands in particular) are believed to affect global climate through the alteration of carbon dioxide and methane cycles. Burning peat as fuel further increases production of greenhouse gases such as carbon dioxide. Current global warming trends are likely to result in increased atmospheric trapping of greenhouse gases, in part because of the release of methane from boreal peat bogs. Wetlands contribute from 33 to 50 percent of the total annual methane production per year (100 teragrams; Whiting & Chanton 1993), mostly from boreal peatlands but approximately 25 percent from tropical and subtropical wetlands as well.

Cultural Services. Recreation such as bird watching, boating, fishing, and hunting are ecosystem services provided by many freshwater food webs that are supported by benthic organisms. In some areas, the recreational catch and value to the economy of recreational fishing outweigh the commercial catch because recreational fishermen spend nearly five times more per fish caught than commercial fishermen (DeSylva 1969). In South America, for example, the Pantanal provides many opportunities for ecotourism and recreational fishing in this enormous tropical wetland (approximately the size of the state of Florida). Its basin includes approximately 138,000 km^2 in Brazil and 100,000 km^2 in Bolivia and Paraguay (see Giller et al., Chapter 6). For four to six months of most years, some 70 percent of the land is inundated. Hunters, fishermen, and conservationists travel from all over the world to view and to exploit this exceptional biodiversity (Moraes & Seidl 1998). During the dry season, this wetland becomes a savanna used for grazing large herds of cattle.

Species Diversity and Ecosystem Services

In many regions, biodiversity is concentrated in specific "hot spots" of high species richness. For example, riparian areas and riverine wetlands typically maintain a much higher biodiversity than the proportion of the landscape that they occupy (National Research Council 2002). Large, ancient ecosystems such as Lake Baikal, the Amazon River, and the Pantanal wetland are other examples of especially diverse biotic communities. The seasonally flooded forests in the Amazon basin contain about 20 percent of the 4,000–5,000 estimated Amazonian tree species (Junk et al. 1989), and fish diversity in the Amazon Basin is exceptionally rich. The Pantanal contains more than 400 species of fish and many species of benthic invertebrates (see Giller et al., Chapter 6) in addition to providing habitat to endangered species such as the giant river otter (*Pteronura braziliensis*).

In the Santa Monica Mountains of Southern California, less than 1 percent of the total land area is comprised of wetlands but approximately 20 percent of the native vascular plant species have their primary habitat there (Rundel & Sturmer 1998). In Sweden, 13 percent (>260 spp) of the country's entire vascular plant flora occur along the Vindel River (Nilsson 1992). In France, 30 percent of the country's 1,386 vascular plant species occur along the Adour (Planty-Tabacchi et al. 1996). Approximately 28 percent of the threatened or endangered plants, 58 percent of threatened or endangered vertebrates and mussels, and 38 percent of threatened or endangered insects in the United States occur in wetlands (Niering 1988). Nevertheless, half of the world's wetlands are estimated to have been lost during the 20th century (Dahl 1990). More than half of this loss has been in the United States, and most resulted from conversion to agriculture and other land uses (Dahl et al. 1991).

Research Needs and Recommendations

Given the many services provided by benthic species living in a wide range of freshwater habitats, we must better understand how to maintain and protect these species and their associated processes. We suggest several areas that need more study to improve management of these critical services:

1. *Link fisheries production to sustainable models of harvest and management that avoid crashes and long-term breakdown of ecosystem functions.* There is an urgent need to strengthen the long-term collection of data on inland fisheries resources if a more complete understanding is to be achieved about the production of benthic invertebrates and determinants of high-quality water. The relationships between safe levels of water quantity and quality that ensure adequate habitats for freshwater species are poorly understood. Too often, minimum values of flow and dissolved oxygen are viewed as sufficient although they are often based on short-term data. In fact, these guidelines do not provide reliable, long-term sustainability. Including margins of error to enhance the "safe minimum levels" will increase reliability and minimize long-term species losses and impairment of benthic ecosystem services.

2. *Communicate results of large-scale, long-term monitoring programs to community-based organizations.* Major changes in water quantity and quality can encourage governmental agencies and local communities to generate alternative actions such as communities that conserve water for sustaining instream flow needs and fisheries that provide sustainable ecosystem services. Results of water-quality monitoring programs need to be translated into formats that enhance effective and informed responses from a wide range of stakeholders. Management groups need to include

wide representation by both professional managers and general consumers of ecosystem services. Community-based ecosystem management approaches will also help to establish systematic data collection on the direct and indirect costs and benefits of fish stocking, introduction, and other "enhancement programs" to determine impacts of nonnative fish species on benthic biodiversity and ecosystem services. Programs such as the European Union's Freshwater Directive, the US Environmental Protection Agency's Community-Based Ecosystem management Program, and the Nature Conservancy's Sustainable Waters Initiative are recent examples of frameworks that are designed to incorporate a wide range of stakeholders in decision making. More of these partnerships and social networks are needed to resolve conflicts regarding evaluations and alternative uses of fresh water.

3. *Monitor and restore habitats in rivers, lakes, and wetlands.* Information on a wider range of chemical and biological measures is needed to detect changes in both surface water and groundwater that are essential to ecosystem services. Enhanced technologies such as remote sensing, wireless data transmission, and comprehensive modeling to develop spatial data are needed to monitor the connections among groundwater levels, stream and river flows, lake-level changes, and wetland distributions at large scales. This regional and cross-national monitoring can provide up-to-date information on changes in the locations, sizes, and types of lakes and wetlands, especially in response to climate changes and global changes in land use. Extreme fluctuations in runoff and erosion from widespread deforestation and related land uses are rapidly altering sedimentary conditions in many fresh waters that, in turn, will alter benthic habitats and associated ecosystem services. Restoration and establishment of hydrological monitoring stations are needed to improve the water-quality monitoring at the regional level using benthic invertebrates and diatoms. Even though there is increased recognition of the long-term effects of climate changes (global warming, cyclic changes in El Nino–Southern Oscillation and the North Atlantic Oscillation, etc.), the capacity to monitor stream flows and lake levels on large-scales within nested watersheds is limited and even diminishing in many regions. Integrated information on data regarding long-term changes in groundwater resources, including their distribution, quality, capacity, and use, is needed across a wide range of scales in different regions.

4. *Restore natural flow regimes.* Information on the number and locations of dams, including the thousands of dams less than 15 m in height that are not currently listed in international databanks, must be compiled and made widely available. These small dams greatly influence the peak flows, minimum low flows, and habitats available for benthic species. Additional studies of the effects of dam removal are needed to identify trade-offs for comparisons with more innovative management of water releases from reservoirs. Many small dams are being removed to provide more upstream habitat for fishes, but sediment releases during and after reservoir removal can still degrade benthic habitats for many years. Furthermore, some nonnative species can increase their distributions following dam removal if the struc-

tures previously served as barriers to dispersal. What ecosystem services are lost when dams are removed? How can these relatively short-term losses be minimized and long-term gains maximized? These and many other questions are being investigated as more dams are being phased out and removed.

5. *Consider additional measures of diversity.* Diversity measures have usually considered just the number of species and/or functional groups in studies of benthic ecosystem processes. This limited approach excludes consideration of the range of diversity elements that potentially affect ecosystem services because different size and age classes within species, as well as their relative abundances, food preferences, and positions within food webs, all can influence rates of processes. Anthropogenic disturbances can substantially change the evenness of species, distributions of abundance, and foraging behavior without associated changes in species richness. The question of whether these changes in evenness, independent of changes in species richness, can influence levels of ecosystem functioning is a necessary focus of future investigation.

Literature Cited

Abramovitz, J.N. 1998. Putting a value on nature's "free" services. *Worldwatch* 11:10–19.

Acharya, G. 2000. Approaches to valuing the hidden hydrological services of wetland ecosystems. *Ecological Economics* 35:63–74. Sp. Iss.

Alexander, R.B., R.A. Smith, and G.E. Schwarz. 2000. Effect of stream channel size on the delivery of nitrogen to the Gulf of Mexico. *Nature* 403:758–761.

Anderson, D.M., P.M. Gilbert, and J.M. Burkholder. 2002. Harmful algal blooms and eutrophication: Nutrient sources, composition and consequences. *Estuaries* 25:704–726.

Baerenklau, K.A., B. Stumborg, and R.C. Bishop. 1999. Nonpoint source pollution and present values: A contingent valuation study of Lake Mendota. *American Journal of Agricultural Economics* 81:1313.

Baker, K.H., and D.S. Herson. 1994. *Bioremediation.* New York, McGraw-Hill.

Barbier, E.B., J.C. Burgess, and C. Folke. 1994. *Paradise Lost? The Ecological Economics of Biodiversity.* London, Earthscan.

Batt, B.D.J., M.G. Anderson, C.D. Anderson, and F.D. Caswell. 1989. The use of prairie potholes by North American ducks. In: *Northern Prairie Wetlands,* edited by A.G. van der Valk, pp. 204–227. Ames, Iowa, Iowa State University Press.

Bedford, B.L. 1996. The need to define hydrological equivalence at the landscape scale for freshwater wetland mitigation. *Ecological Applications* 6:57–68.

Bedford, B.L. 1999. Cumulative effects on wetland landscapes: Links to wetland restoration in the United States and southern Canada. *Wetlands* 19:775–788.

Bockstael, N.E., A.M. Freeman, R.J. Kopp, P.R. Portney, and V.K. Smith. 2000. On measuring economic values for nature. *Environmental Science and Technology* 34:1384–1389.

Bonilla-Warford, C., and J.B. Zedler. 2002. Potential for using native plant species in stormwater wetlands. *Environmental Management* 29:385–393.

Bouchard, D.C., M.K. Williams, and R.Y. Surampalling. 1992. Nitrate contamination of groundwater sources and potential health effects. *Journal of American Water Works Association* 84:58–90.

Boyle, K.J., P.J. Poor, and L.O. Taylor. 1999. Estimating the demand for protecting freshwater lakes from eutrophication. *American Journal of Agricultural Economics* 81:1118–1122.

Brinson, M.M., and A.I. Malvarez. 2002. Temperate freshwater wetlands: Types, status and threats. *Environmental Conservation* 29:115–133.

Brismar, A. 2002. River systems as providers of goods and services: A basis for comparing desire and undesired effects of large dam projects. *Environmental Management* 29:598–609.

Brock, T.D. 1985. *A Eutrophic Lake: Lake Mendota, Wisconsin.* New York, Springer-Verlag.

Bronmark, C., and L.A. Hansson. 1998. *The Biology of Lakes and Ponds.* New York, Oxford University Press.

Burkholder, J.M. 1998. Implications of harmful microalgae and heterotrophic dinoflagellates in management of sustainable marine fisheries. *Ecological Applications* 8:S37–S62.

Carpenter, S.R., D. Ludwig, and W.A. Brock. 1999. Management and eutrophication for lakes subject to potentially irreversible change. *Ecological Applications* 9:751–771.

Carpenter, S.R., and M. Turner. 2000. Opening the black boxes: Ecosystem science and economic valuation. *Ecosystems* 3:1–3.

Cioc, M. 2002. *The Rhine: An Eco-biography, 1815–2000.* Seattle, University of Washington Press.

Clements, W.H., and M.C. Newman. 2002. *Community Ecotoxicology.* New York, John Wiley and Sons.

Cleveland, C.J., D.I. Stern, and R. Costanza. 2001. *The Economics of Nature and the Nature of Economics.* Cheltenham, UK and Northampton, Massachusetts, Edward Elgar Publishers.

Costa-Pierce, B.A. 1998. Preliminary investigation of an integrated aquaculture-wetland ecosystem using tertiary-treated municipal wastewater in Los Angeles County, California. *Ecological Engineering* 10:341–354.

Covich, A.P., M.C. Austen, F. Bärlocher, E. Chauvet, B.J. Cardinale, C.L. Biles, P. Inchausti, O. Dangles, M. Solan, M.O. Gessner, B. Statzner, B. Moss, 2004. The role of biodiversity in the functioning of freshwater and marine benthic ecosystems. *BioScience* 54:767–775.

Covich, A.P., M.A. Palmer, and T.A. Crowl. 1999. The role of benthic invertebrate species in freshwater ecosystems. *BioScience* 49:119–127.

Crowl, T.A., W.H. McDowell, A.P. Covich, S.L. Johnson. 2001. Freshwater shrimp effects on detrital processing and localized nutrient dynamics in a montane, tropical rain forest stream. *Ecology* 82:775–783.

Cuffney, T.F., J.B. Wallace, and G.J. Lugthart. 1990. Experimental evidence quantifying the role of benthic invertebrates in organic matter dynamics of headwater streams. *Freshwater Biology* 23:281–299.

Dahl, T.E. 1990. *Wetland Losses in the United States, 1789 to 1980.* Washington, DC, US Department of the Interior, Fish and Wildlife Service.

Dahl, T.E., C.E. Johnson, and W.E. Frayer. 1991. *Status and Trends of the Wetlands in the Conterminous United States Mid-1970s to Mid-1980s.* Washington, DC, US Department of the Interior, Fish and Wildlife Service.

Dahm, C.N., K.W. Cummins, H.M. Valet, and R.L. Coleman. 1995. An ecosystem view of the restoration of the Kissimmee River. *Restoration Ecology* 3:225–238.

Daily, G.C., editor. 1997. *Nature's Services: Societal Dependence on Natural Ecosystems.* Washington, DC, Island Press.

Daily, G.C., and K. Ellison. 2002. New York: How to put a watershed to work. In: *The New Economy of Nature,* edited by G.C. Daily and K. Ellison, pp. 61–85. Washington, DC, Island Press.

Davies, B., and J. Day. 1998. *Vanishing Waters.* Cape Town, South Africa, University of Cape Town Press.

DeBruyn, A.M.H., and J.B. Rasmussen. 2002. Quantifying assimilation of sewage-derived organic matter by riverine benthos. *Ecological Applications* 12:511–520.

DeSylva, D.P. 1969. Trends in marine sport fisheries research. *American Fisheries Society Transactions* 98:151–169.

Dobson, M., A. Magana, J.M. Mathooko, and F.K. Ndegwa. 2002. Detritivores in Kenyan highland streams: More evidence for the paucity of shredders in the tropics? *Freshwater Biology* 47: 909–919.

Dudgeon, D. 2003. Clinging to the wreckage: Unexpected persistence of freshwater biodiversity in a degraded tropical landscape. *Aquatic Conservation: Marine and Freshwater Ecosystems* 13:93–97.

Everard, M., and A. Powell. 2002. Rivers as living systems. *Aquatic Conservation: Marine and Freshwater Ecosystems* 12: 329–337.

Ewel, K.C. 1997. Water quality improvement by wetlands. In: *Nature's Services: Societal Dependence on Natural Ecosystems,* edited by G.C. Daily, pp. 329–344. Washington, DC, Island Press.

FAO. 1995. Food and Agricultural Organization of the United Nations. *Review of the State of World Fishery Resources: Inland Capture Fisheries.* Rome, Italy: FAO Fisheries Circular #885.

FAO. 2003. Food and Agriculture Organization of the United Nations, Rome, Italy (http://www.fao.org/).

Foran, J., T. Brosnan, M. Connor, J. Delfino, J. DePinto, K. Dickson, H. Humphrey, V. Novotny, R. Smith, M. Sobsey, and S. Stehman. 2000. A framework for comprehensive, integrated, waters monitoring in New York City. *Environmental Monitoring and Assessment* 62:147–167.

Gandy, M. 2002. *Concrete and Clay: Reworking Nature in New York City.* Cambridge, Massachusetts, MIT Press.

Geber, U., and J. Bjorklund. 2001. The relationship between ecosystem services and purchased input in Swedish wastewater treatment systems: A case study. *Ecological Engineering* 18:39–59.

Ghiorse, W.C., and J.T. Wilson. 1988. Microbial ecology of the terrestrial subsurface. *Advances in Applied Microbiology* 33:107–172.

Giller, P.S., H. Hillebrand, U.-G. Berninger, M. Gessner, S. Hawkins, P. Inchausti, C. Inglis, H. Leslie, B. Malmqvist, M. Monaghan, P. Morin, and G. O'Mullan. 2004. Biodiversity effects on ecosystem functioning: Emerging issues and their experimental test in aquatic communities. *Oikos* 104:423–436.

Giller, P.S., and B. Malmqvist. 1998. *The Biology of Streams and Rivers.* Oxford, Oxford University Press.

Gounot, A.M. 1994. Microbial ecology of groundwaters. In: *Groundwater Ecology,* edited by J. Gibert, D.L. Danielopol, and J.A. Stanford, pp. 189–215. San Diego, California, Academic Press.

Gupta, S.K., R.C. Gupta, A.K. Seth, A.B. Gupta, J.K. Bassin, and A. Gupta. 2000. Methaemoglobinaemia in areas with high nitrate concentrations in drinking water. *National Medical Journal of India* 13:58–61.

Hart, J. 1996. *Storm Over Mono: The Mono Lake Battle and the California Water Future.* Berkeley, California, University of California Press.

Heal, G. 2000.Valuing ecosystem services. *Ecosystems* 3:24–30.

Heard, S.B., and J.S. Richardson. 1995. Shredder-collector facilitation in stream detrital food webs: Is there enough evidence? *Oikos* 72:359–366.

Hobbs, H.H., III. 2001. Decapoda. In: *Ecology and Classification of North American Freshwater Invertebrates,* edited by J.H. Thorp and A.P. Covich, pp. 955–1001. San Diego, California, Academic Press.

Huner, J.V. 1995. An Overview of the status of freshwater crawfish culture. *Journal of Shellfish Research* 14:539–543.

IRRI. 2000. International Rice Research Institute Statistics, Manila, Philippines (http://www.irri.org/science/ricestat/index.asp).

Jasinski, S.M. 1999. Peat. In: *Minerals Yearbook 1999: Volume I: Metals and Minerals. Minerals and Information.* http://minerals.usgs.gov/minerals/pubs/commodity/peat/510499.pdf. Reston, Virginia, US Geological Survey.

Johnson, R.K., T. Wiederholm, and D.M. Rosenberg. 1992. Freshwater biomonitoring using individual organisms, populations, and species assemblages of benthic macroinvertebrates. In: *Freshwater Biomonitoring and Benthic Macroinvertebrates,* edited by D.M. Rosenberg and V.H. Resh, pp. 40–158. New York, Chapman and Hall.

Jonsson, M., and B. Malmqvist. 2003. Importance of species identity and number for process rates within different functional feeding groups. *Journal of Animal Ecology* 72:453–459.

Junk, W.J., P.B. Bailey, and R.E. Sparks. 1989. The flood pulse concept in river-floodplain systems. *Canadian Special Publication Fisheries and Aquatic Sciences* 106:110–127.

Kadlec, R.H., and R.L. Knight. 1996. *Treatment Wetlands.* Boca Raton, Florida, Lewis Publishers.

Keddy, P.A. 2000. *Wetland Ecology: Principles and Conservation.* Cambridge, UK, Cambridge University Press.

Loomis, J. 1987. Balancing public trust resources of Mono Lake and Los Angeles' water right: An economic approach. *Water Resources Research* 23:1449–1456.

Loomis, J. 1995. Public trust doctrine produces water for Mono Lake. *Journal of Soil and Water Conservation* 50:270–271.

Loomis, J.B. 2000. Environmental valuation techniques in water resource decision making. *Journal of Water Resources Planning and Management* 126:339–344.

Loomis, J., P. Kent, L. Strange, K. Fausch, A. Covich. 2000. Measuring the total economic value of restoring ecosystem services in an impaired river basin: Results from a contingent valuation survey. *Ecological Economics* 33:103–117.

Lugthart, G.J., and J.B. Wallace. 1992. Effects of disturbance on benthic functional structure and production in mountain streams. *Journal of the North American Benthological Society* 11:138–164.

Mallin, M.A. 2000. Impacts of industrial animal production on rivers and estuaries. *American Scientist* 88:2–13.

Malmqvist, B., and S. Rundle. 2002. Threats to the running water ecosystems of the world. *Environmental Conservation* 29:134–153.

Marmonier, P., P. Vervier, J. Gibert, and M.-J. Dole-Olivier. 1993. Biodiversity in ground waters: A research field in progress. *Trends in Ecology and Evolution* 8:392–395.

Mattison R.G., and S. Harayama. 2001. The predatory soil flagellate *Heteromita globosa* stimulates toluene biodegradation by a *Pseudomonas* sp. *FEMS Microbiology Letters* 194:39–45.

Mayer, P.M., and S.M. Galatowitsch. 2001. Assessing ecosystem integrity of restored prairie wetlands from species production-diversity relationships. *Hydrobiologia* 443:177–185.

Meyer, J.L., and G.E. Likens. 1979. Transport and transformations of phosphorus in a forest stream ecosystem. *Ecology* 60:1255–1269.

Millennium Ecosystem Assessment. 2003. *Ecosystems and Human Well-Being: A Framework for Assessment.* Washington, DC, Island Press.

Mitsch, W.J., and J.G. Gosselink. 1993. Values and valuation of wetlands. In: *Wetlands, Second Edition,* edited by W.J. Mitsch and J.G. Gosselink, pp. 507–540. New York, Van Nostrand Reinhold.

Mitsch, W.J., and J.G. Gosselink. 2000. The value of wetlands: Importance of scale and landscape setting. *Ecological Economics* 35:25–33. Sp. Iss.

Moeszlacher, F. 2000. Sensitivity of groundwater and surface water crustaceans to chemical pollutants and hypoxia: Implications for pollution management. *Archiv für Hydrobiologie* 149:51–66.

Moraes, A.S., and A.F. Seidl. 1998. Sport fishing trips to the southern Pantanal (Brazil). *Brazilian Review of Agricultural Economics and Rural Sociology* 36:211–226.

Moshiri, G.A., editor. 1993. *Constructed Wetlands for Water Quality Improvement.* Boca Raton, Florida, CRC Press.

Moulton, T.P. 1999. Biodiversity and ecosystem functioning in conservation of rivers and streams. *Aquatic Conservation: Marine and Freshwater* 9:573–578.

Naiman, R.J., and H. Decamps. 1997. The ecology of interfaces: Riparian zones. *Annual Review of Ecology and Systematics* 28:621–658.

National Research Council. 2002. *Riparian Areas: Functions and Strategies for Management.* Washington, DC, National Academy Press.

National Research Council. In press. Translating Ecosystem Functions to the Value of Ecosystem Services. In: *Ecosystem Services of Freshwater and Associated Ecosystems.* Washington, DC, National Research Council Press.

New, M.B., and W.C. Valenti. 2000. *Freshwater Prawn Culture.* Oxford, Blackwell Science.

Niering, W.A. 1988. Endangered, threatened and rare wetland plants and animals of the continental United States. In: *The Ecology and Management of Wetlands,* edited by D.D. Hook, W.H. McKee, H.K. Smith, J. Gregory, V.G. Burrell, Jr., M.R. DeVoe, R.E. Sojka, S. Gilbert, R. Banks, L.H. Stolzy, C. Brooks, T.D. Matthews, and T.H. Shear, p. 592. Portland, Oregon, Timber Press.

Nilsson, C. 1992. Conservation and management of riparian communities. In: *Ecological Principles of Nature Conservation,* edited by L. Hansson, pp. 352–372. London, Elsevier.

Nilsson, C. 1999. Rivers and streams. In *Swedish Plant Geography,* edited by H. Rydin, P. Snoeijs, and M. Diekmann, Vol. 84 pp. 135–148 Acta Phytogeograqphica Suecica. Uppsala, Sweden, Svenska Växtgeografiska Sällskapet.

Novak, M., J.A. Balen, M.E. Obbard, and B. Mallocheds (eds.). 1987. *Wildlife Furbearer Management and Conservation in North America.* Ontario, Canada, Ontario Trappers Association.

Odum, H.T. 1984. Summary: Cypress swamps and their regional role. In: *Cypress Swamps,* edited by K.C. Ewel and H.T. Odum, pp. 416–444. Gainesville, Florida, University Press of Florida.

Odum, H.T., and E.P. Odum. 2000. The energetic basis for valuation of ecosystem services. *Ecosystems* 3:21–23.

O'Melia, C.R., M.J. Pfeffer, P.K. Barten, G.E. Dickey, M.W. Garcia, C.N. Haas, R.G. Hunter, R.R. Lowrance, C.L. Moe, C.L. Paulson, R.H. Platt, J.L. Schnoor, T.R. Schueler, J.M. Symons, and R.G. Wetzel. 2000. *Watershed Management for Potable Water Supply. Assessing the New York City Strategy.* Washington, DC, National Research Council, National Academy Press.

Palmer, M.A., A.P. Covich, S. Lake, P. Biro, J.J. Brooks, J. Cole, C. Dahm, J. Gibert, W. Goedkoop, J. Verhoeven, and W.J. Van De Bund. 2000. Linkages between aquatic sediment biota and life above sediments as potential drivers of biodiversity and ecological processes. *BioScience* 50:1062–1075.

Pearce, D. 1998. Auditing the earth: The value of the world's ecosystem services and natural capital. *Environment* 40:23–28.

Pind, A., N. Risgaard-Petersen, and N.P. Revsbech. 1997. Denitrification and microphytobenthic NO_3^- consumption in a Danish lowland stream: Diurnal and seasonal variation. *Aquatic Microbial Ecology* 12:275–284.

Planty-Tabacchi, A.M., E. Tabacchi, R.J. Naiman, C. DeFerrari, and H. Decamps. 1996. Invisibility of species-rich communities in riparian zones. *Conservation Biology* 10:598–607.

Postel, S.L., and S.R. Carpenter. 1997. Freshwater ecosystem services. In: *Nature's Services: Societal Dependence on Natural Ecosystems,* edited by G.C. Daily, pp. 195–214. Washington, DC, Island Press.

Rosén, E., and S. Borgegård. 1999. The open cultural landscape. In: *Swedish Plant Geography,* edited by H. Rydin, P. Snoeijs, and M. Diekmann, Vol. 84 pp. 113–134. Acta Phytogeograqphica Suecica. Uppsala, Sweden, Svenska Växtgeografiska Sällskapet.

Rundel, P.W., and S.B. Sturmer. 1998. Native plant diversity in riparian plant communities of the Santa Monica Mountains, California. *Maroño* 45:93–100.

Rydin, H., H. Sjörs, and M. Löfroth. 1999. Mires. In: *Swedish Plant Geography,* edited by H. Rydin, P. Snoeijs, and M. Diekmann, Vol. 84 pp. 91–112 Acta Phytogeograqphica Suecica. Uppsala, Sweden, Svenska Växtgeografiska Sällskapet.

Stevenson, R.J., and F.R. Hauer. 2002. Integrating hydrogeomorphic and Index of Biotic Integrity approaches for environmental assessment of wetlands. *Journal of the North American Benthological Society* 21:502–513.

Strange, E., K.D. Fausch, and A.P. Covich. 1999. Sustaining ecosystem services in human-dominated watersheds: Biohydrology and ecosystem processes in the South Platte River Basin. *Environmental Management* 24:39–54.

Sweeney, B. 2003. Personal communication.

Taft, O.W., M.A. Colwell, C.R. Isola, and R.J. Safran. 2002. Waterbird responses to experimental drawdown: Implications for the multispecies management of wetland mosaics. *Journal of Applied Ecology* 39:987–1001.

Thorp, J.H., and A.P. Covich (eds.). 2001. *Ecology and Classification of North American Freshwater Invertebrates.* San Diego, California, Academic Press.

Toth, L.A., S.L. Melvin, D.A. Arrington, and J. Chamberlain. 1998. Hydrologic manipulations of the channelized Kissimmee River: Implications for restoration. *BioScience* 48:757–764.

Usui, M., Y. Kakuda, and P.G. Kevan. 1994. Composition and energy values of wild fruits from the Boreal Forest of northern Ontario. *Canadian Journal of Plant Science* 74:581–587.

Wall, D.H., P.V.R. Snelgrove, and A.P. Covich. 2001. Conservation priorities for soil and sediment invertebrates. In: *Conservation Biology,* edited by M.E. Soulé and G.H. Orians. Washington, DC, Island Press.

Wallace, J.B., J.W. Grubaugh, and M.R. Whiles. 1996. Biotic indices and stream ecosystem processes: Results from an experimental study. *Ecological Applications* 6:140–151.

Wallenius, T.H. 1999. Yield variations of some common wild berries in Finland in 1956–1996. *Annales Botanici Fennici* 36:299–314.

Watson, S.B., and J. Lawrence. 2003. Drinking water quality and sustainability. *Water Quality Research Journal of Canada* 38:3–13.

Welcomme, R.L. 2001. *Inland Fisheries: Ecology and Management.* Oxford, Blackwell Science.

Weller, M.W. 1981. *Freshwater Marshes.* Minneapolis, Minnesota, University of Minnesota Press.

Whalen, P.J., L.A. Toth, J.W. Koebel, and P.K. Strayer. 2002. Kissimmee River restoration: A case study. *Water Science and Technology* 45:55–62.

Wharton, C.H., W.M. Kitchens, E.C. Pendelton, and T.W. Snipe. 1982. *The Ecology of Bottomland Hardwood Swamps of the Southeast: A Community Profile.* FWS/OBS-81/37. Washington, DC, US Fish and Wildlife Service.

Whigham, D.F. 1997. Ecosystem functions and ecosystem values. In: *Ecosystem Function and Human Activities: Reconciling Economics and Ecology,* edited by R.D. Simpson and N.L. Christensen, pp. 225–239. New York, Chapman and Hall.

Whiting, G.J., and J.P. Chanton. 1993. Primary production control of methane emission from wetlands. *Nature* 364:794–795.

Williams, W.D. 1999. Conservation of wetlands in drylands: A key global issue. *Aquatic Conservation: Marine and Freshwater* 9:517–522.

Williams, W.D. 2002. Environmental threats to salt lakes and the likely status of inland saline ecosystems in 2025. *Environmental Conservation* 29:154–167.

Zedler, J.B. 2000. Progress in wetland restoration ecology. *Trends in Ecology and Evolution* 15:402–407.

Zelenke, J. 1998. Tidal freshwater marshes of the Hudson River as nutrient sinks: Long-term retention and denitrification. Final Report to the Hudson River Foundation, Inc. GF/03/96, Cambridge, MD, University of Maryland Center for Environmental Science.

4

Marine Sedimentary Biota as Providers of Ecosystem Goods and Services

Jan Marcin Weslawski, Paul V.R. Snelgrove, Lisa A. Levin, Melanie C. Austen, Ronald T. Kneib, Thomas M. Iliffe, James R. Garey, Stephen J. Hawkins, and Robert B. Whitlatch

Marine sediments cover more of the Earth's surface than all other ecosystems combined (Snelgrove 1999), yet direct human experience is limited largely to the narrow zone at the interface between land and sea. Although 62 percent of the Earth's surface is covered by water greater than 1,000 m deep, only approximately 2 km^2 (Paterson 1993) has been quantitatively sampled for macrofauna (invertebrates greater than 300 microns but not identifiable in photographs) and only 5 m^2 (Lambshead 1993) has been sampled for meiofauna (invertebrates greater than 300 microns but retained on a 44-micron sieve). With most of the ocean sedimentary biota out of sight, we tend to ignore their role in regulating rates and processes that maintain the integrity of marine systems (Snelgrove et al. 1997), instead focusing on biologically generated products or consequences that are of direct economic benefit. The publicity associated with the Kyoto Protocol (United Nations 1992), particularly with respect to atmospheric carbon dioxide increases and carbon sequestration, has helped to broaden public concern about the role of the sea in climate regulation, but even here, the primary focus has been on the water column above the seafloor and its processes (Martin et al. 1994; Hanson et al. 2000). Public outcry in the United States and elsewhere has driven major changes in environmental policy over the last 20 years, resulting in significant improvement in environmental standards for air, land, and drinking water, and improved protection for species that are considered endangered. Unfortunately, oceans have not received similar levels of protection. Seaward deposition of waste materials generated in the terrestrial domain continues generally without regard for effects on sediments and marine benthos. Most marine sedimentary organisms are undescribed (Grassle & Maciolek 1992) and have no degree of protection.

Table 4.1. Goods and services provided by sedimentary systems.

The role of sedimentary invertebrates has been inferred from published studies; ratings are based on estimated global importance. Public concern is based on qualitative observations of how frequently the good or service is discussed in the popular press.

	Role of Sedimentary Biota	*Public Concern*
Provisioning		
Animal food	moderate	high
Plant food	low	low
Medicine & models for human research	moderate	low
Fuels & energy	high (on geological time scales)	low
Clean water	high	moderate
Fiber	low	low
Regulating Services		
Remineralization	high	low
Waste treatment	high	low
Biological control	moderate	low
Gas and climate regulation	moderate	moderate
Disturbance regulation	moderate	low
Erosion and sedimentation control	high	low
Habitat Maintenance Services		
Landscape linkage & structure/ habitat/refugia	high	moderate
Aesthetic Services		
Recreation, tourism, and education	high	high

Although oceans are responsible for approximately 60 percent of the estimated total value of global ecosystem services (Costanza 1999), efforts to valuate the specific roles of sedimentary biota are effectively nonexistent.

In this chapter we identify ecosystem processes that are strongly influenced or regulated by marine sedimentary systems, and consider how marine sedimentary organisms contribute to economically important extractable *ecosystem goods/products* (e.g., fish) and influence *ecosystem services* (e.g., water purification and shoreline stabilization, see Chapter 1, Figure 1.1) within the marine environment. We include a summary of important ecosystem goods and services provided by marine sedimentary biota (Table 4.1), the roles that living organisms play in delivering those goods and services, the biotic and abiotic factors that regulate provisioning of services, and specifically how biodiversity contributes to regulation and provisioning of ecosystem goods and services. The marine systems considered here are grouped into estuarine, continental shelf, and deep-sea sediments. *Estuaries* encompass sedimentary habitat at the land-sea interface where freshwater input measurably dilutes seawater, *continental shelf sediments* refer to the submerged, gently sloping seafloor between continents and the upper edge of the continental slope (~130 m deep), and *deep-sea sediments* include the comparatively steep (~4°) continental slope that extends from the edge of the continental shelf to the less steep continental rise (~4,000 m) that grades into the abyssal plains (4,000–6,000 m). Abyssal plains are primarily sediment-covered, flat rolling plains that cover approximately 40 percent of the Earth's surface; in some areas they contain submerged mountains known as *seamounts* that can extend thousands of meters above the seafloor to relatively shallow depths. Threats and sustainability of goods and services in these habitats are addressed in Chapter 7.

Estuarine and Continental Shelf Sediments

Approximately 39 percent of the global human population, or approximately 2.2 billion people, lived within 100 km of the coast in 1995, most within estuarine watersheds (Burke et al. 2001). In countries such as the United States, coastal populations have increased faster than the overall population (Beach 2002). Historically, human populations have depended on estuaries for food (e.g., fish and shellfish), transportation, trade (e.g., waterways, sheltered ports), and recreation. Ancient civilizations in the Fertile Crescent (area around the rivers Tigris, Euphrates, Nile, and on the western slopes of the Mediterranean coast) are now recognized to have had a culture and society that were based on utilization of wetlands and estuaries (Pournelle 2003). This dependence on estuaries has arisen because these sedimentary environments harbor abundant fishes and shellfishes, are habitats for many invertebrates that are also integral parts of estuarine and oceanic food webs, and are essential for the long-term sustainability of coastal ecosystems.

Wherever they occur, vascular plants contribute to virtually every ecosystem service associated with estuaries. Although restricted to intertidal (e.g., marshes and mangroves) and shallow subtidal (seagrass beds) portions of temperate and tropical estuaries, the contribution of these plant communities to estuarine production can be greater than suggested by their modest areal extent (Heymans & Baird 1995). Aboveground plant structures (e.g., stems and leaves of marsh plants or prop roots of mangroves) trap and retain sediments, and provide substrata, refugia, and food for estuarine biota (Thayer et al. 1987; Covi & Kneib 1995). Plant roots help to stabilize sediments and promote the structural integrity of tidal channels, and mediate biological activity in the sediments by transporting oxygen to the root zone and detoxifying sediments (Lee et al. 1999). Benthic plants and animals also maintain environmental quality by binding and removing particulates and contaminants from the water column and sediments and are an integral part of the aesthetic vistas of coastal landscapes that enrich the human spirit.

Although estuarine sediments contain few species relative to most other sedimentary habitats, they nonetheless represent hotspots for ecosystem processes that can extend well beyond the estuarine sediments. Of the ecosystem goods and services associated with shelf and nearshore ocean areas, people are most aware of provisioning of food (e.g., fish and shellfish), which has huge commercial and cultural importance in coastal societies worldwide. Even aquaculture businesses often rely on wild (natural) fisheries (e.g., for fishmeal) or natural supply of food (e.g., phytoplankton) for aquaculture species and, in some cases, for provision of brood and juvenile stocks. Marine plants are used as food, particularly in Asia, and seaweed extracts such as alginates and other phycocolloids are used in many industrial and food applications (e.g., manufacture of films, rubber, linoleum, cosmetics, paints, cheeses, lotions). The living components of estuarine systems provide not only the primary and secondary production that supports commercial, recreational, and subsistence fishing and other extractable resources, but also much of the structure that stabilizes sediments to provide flood and erosion control, and maintains the integrity of wetlands and coastal waterways (Levin et al. 2001a; Tables 4.2a–4.2b).

Sedimentary fauna are a critical part of the diet for many estuarine and shelf species that feed near or on the bottom, such as cod and flatfish (Feder & Pearson 1988; Carlson et al. 1997). Some pelagic fish feed directly on benthic invertebrates at the seafloor-water interface during various phases in their life cycles. Many benthic fauna spend the early parts of their life cycle in the plankton and, in some cases, are extremely abundant and potentially important for pelagic food chains (Lindley et al. 1995). Structure-rich sedimentary habitats, particularly marshes, mangrove swamps, and seagrass beds, create refuges for juveniles of commercially exploited pelagic fish and invertebrates (Laurel et al. 2003).

Nutrient cycling and sediment oxygenation (redox) processes are interlinked to lesser known, but key, services of detoxification and disposal of waste by shelf and estuarine sediment biota. These processes are regulated directly by microbial organisms and indirectly by larger, bioturbating organisms (Henriksen et al. 1983; Pelegri & Blackburn 1995). Detoxification and immobilization of contaminants may represent a service or a disservice, depending on the circumstances. Detoxification is performed primarily by microbes (Geiselbrecht et al. 1996) and may be facilitated by bioturbation, which strongly influences oxygenation and physical movement of contaminants. Bioturbating organisms such as polychaete worms relocate sediment particles and water as they feed, and amalgamate fine particles into fecal pellets (Levinton 1995). Microbes process organic wastes and organic compounds into less hazardous breakdown products (Boyd & Carlucci 1996; Lee & Page 1997), which can be recirculated back into the water column through bioturbation. Microbial processing of toxic waste such as organometallic compounds can produce harmful breakdown products that can be biomagnified through the food web (Srinivasan & Mahajan 1989). Bioturbation activity by large invertebrates can also accelerate pollutant burial by feeding and removing material at the sediment surface and defecating deeper in the sediment, but feeding at depth by other species that defecate at the surface can also remobilize buried contaminants (Gallagher & Keay 1998).

Sediment-dwelling organisms contribute to sediment formation through their skeletal remains (e.g., the shells and calcareous structures of mollusks, foraminifera, and lithothamnia [algae]). More importantly, particularly in nearshore, shallow-subtidal habitats, sedimentary organisms directly affect sediment stability and erodability (Levinton 1995; Paterson & Black 1999). Sediment particles are bound together by extracellular polymeric substances (mucus) within diatom and microbial films (Grant & Gust 1987), and within meiofaunal and macrofaunal secretions. Macrofaunal fecal and pseudofecal production also binds sediments (Rhoads 1963). Although biological adhesion (Grant et al. 1982) and biological structures above the sediment (such as seagrass, Fonseca & Fisher 1986), can stabilize sediment, biologically generated bottom roughness (Wright et al. 1997) and increased water content of sediments as a result of bioturbation (Rhoads & Young 1970) can also increase erodability.

Shelf and estuarine sediments are habitats for many fishes and invertebrates, and are valued for recreation, sport and subsistence fishing. Sandy beaches, for example, are of particular importance as recreational areas (Weslawski et al. 2000). Sediments provide educational value because of their role in the ecosystem and can have spiritual importance for humans as a source of food, ornaments, and even currency (shells).

Estuaries are the most accessible marine sedimentary habitats for humans, and they are also the most productive. The value of ecological services from estuaries can be substantial, an observation that can be attributed to the service of nutrient cycling defined as the storage, internal cycling, processing, and acquisition of nutrients (Costanza et al. 1997; Ewel et al. 2001). In open estuaries, much of the nutrient cycling occurs in the water column, but the benthic component in shallow subtidal and intertidal systems is also important. As with estuaries, depending on the local communities' values and willingness to pay (Daily et al. 2000; Dasgupta et al. 2000), the value of ecosystem services for intertidal wetlands could be substantial. Intertidal wetlands provide critical services such as waste treatment, environmental buffering/flood control, recreation, and food production. Many service categories (e.g., nutrient cycling) must be considered based on their value at local levels; thus, total economic value of these systems may be underestimated at regional and global levels. It is also important to recognize that many methods have been applied in placing monetary values on estuarine habitats, including the substantial cost of restoration to recover lost functionality (Kruczynski 1999). There is insufficient evidence available to know whether estuaries can be restored to all previous functions, although partial restoration of some functions has been achieved in some cases (see Snelgrove et al. Chapter 7).

Continued on page 83

Table 4.2a. The provisioning of goods and services for estuaries.

We have used a qualitative ranking scale from −3 to +3 to compare the relative importance of a given good or service ("Rank") within estuaries. Negative scores denote situations where sedimentary fauna can negatively influence a good or process (e.g., remobilizing pollutants into the environment). We have also estimated the relative importance of species, functional, and habitat diversity in the delivery of a given good or service ("Diversity Importance") using a relative scale from 0 to 3. These rankings are qualitative and largely based on inference rather than diversity studies *per se*. Where information was insufficient to allow assignment of a rank value, a question mark was entered in the table and a value of zero was used in sums.

	Rank	Biotic Contributors	Abiotic Regulators	External Interaction	Diversity Importance		
					Species	Functional	Habitat
Provisioning Services							
Plants as food	−1 to 1	grazers, pathogens	oxygen, circulation, substrate, nutrients		1	1	1.5
Animals as food	−3 to 3	fish, invertebrates, all zoo- and phyto-benthos, pathogens	oxygen, circulation, substrate	food, life history, detritus	3	3	3
Other biological products	−3 to 3	bait worms	sediment type		1	1	2
Biochemical/medicine/models for human research	2	microbes, natural products, invertebrate models	temperature, chemical availability, sediment type		3	3	3
Fuels/energy	1	microbes, peat, mangroves	temperature, time		1	1	1
Fiber	1	sponges, sea grasses, mangroves	water flow, sediment type		1	1	2
C sequestration	1	microbes, peat	sediment type, redox		1	2	2
Nonliving materials (geological effects)	2	bioturbators, microbes, infauna, maerl, shellgravels	hydrodynamic processes, sediment type		1	1	2
Clean seawater	3	seagrasses, saltmarsh plants, biofiltration, bioturbators	hydrodynamic processes, redox, sediment type		2	3	3

Regulation Services

Sediment formation: biodeposition	−1 to 3	microbes, lithothamnia, biogenic sediments, vegetation, filter feeders, infauna	hydrodynamic processes, freshwater & land runoff, sediment type	sinking of particulates	1	2	3
Nutrient cycling	3	microbes, bioturbators, macrofauna, fishes, phytobenthos	hydrodynamic processes, temperature	resuspension	3	3	3
Biological control: disease, invasive species resistance	?	?	oxygen, eutrophication, sedimentation, salinity		3	3	3
Detoxification, waste disposal	−3 to 3	microbes, zoo- & phytobenthos, biofilters & bioturbators	circulation, resuspension, sedimentation		3	3	3
Climate regulation (C sequestration)	1	bioturbators, microbes, infauna, mobile fauna	hydrodynamic processes, upwelling, resuspension, sedimentation	terrestrial & pelagic input	1	1	3
Food web support processes	3	entire benthos	hydrodynamic processes, upwelling, resuspension, sedimentation, oxygen	terrestrial & pelagic input	3	3	3
Atmosphere composition	2	microbes, kelps, wetlands	oxygen, substrate, turbulent mixing, wind	?	1	1	2
Flood and erosion control	3	vascular plants, biostabilizers			1	2	2
Redox processes	3	bioturbators, microbes	oxygen	carbon flux	3	2	2

(continued)

Table 4.2a. *(continued)*

	Rank	Biotic Contributors	Abiotic Regulators	External Interaction	Diversity Importance Species	Functional	Habitat
Habitat Maintenance Services							
Landscape linkages & structure/ habitat/refugia	3	vegetation, biogenic reefs, migrating fauna	oxygen, temperature, depth, substrate	carbon flux, larval stages	2	3	3
Aesthetic Services							
Spiritual/cultural	3				3	1	3
Aesthetic	3				3	1	3
Recreation	3				2	1	3
Scientific understanding	3	ecological paradigms & education			3	3	3

Table 4.2b. The provisioning of goods and services for shelf sediment ecosystems.

See Table 4.2a for explanation of ranking scheme.

	Rank	Biotic Contributors	Abiotic Regulators	External Interaction	Diversity Importance		
					Species	Functional	Habitat
Provisioning Services							
Plants as food	0						
Animals as food	3	fish, invertebrates, all benthos	oxygen, circulation, substrate	food, life history	3	3	3
Other biological products	0						
Biochemical/medicines/ models for human research	2	microbes, natural products, enzymes	temperature, chemical availability		2	2	2
Fuels/energy	3	microbes	temperature, time		1	1	0
Fiber	1	Sponges			1	1	2
C sequestration	1	bioturbators, microbes, infauna	CO_2, temperature, advection	carbon pump	1	2	1
Nonliving materials (geological effects)	0						
Clean seawater	1				1	1	1
Regulation Services							
Sediment formation: biodeposition	2	microbes, lithothamnia, biogenic sediments	currents, freshwater and land runoff	sinking of particulates	2	2	2
Nutrient cycling	3	microbes, bioturbators, macrofauna, fishes	circulation, temperature, tides	resuspension	2	3	2
Biological control: disease, invasive species resistance	?				?	?	?

(continued)

Table 4.2b. (continued)

	Rank	Biotic Contributors	Abiotic Regulators	External Interaction	Diversity Importance		
					Species	Functional	Habitat
Detoxification, waste disposal	−3 to 3	microbes, benthos	circulation, resuspension, sedimentation		3	3	3
Climate regulation (C sequestration)	1	bioturbators, microbes, infauna, mobile fauna	hydrodynamic processes, upwelling, resuspension, sedimentation	terrestrial & pelagic input	2	2	2
Food web support processes	3	entire benthos	hydrodynamic processes, upwelling, resuspension, sedimentation, oxygen	terrestrial & pelagic input	3	3	3
Atmosphere composition	1	microbes	oxygen, substrate, turbulent mixing		1	1	1
Flood and erosion control	1				1	2	2
Redox processes	3	bioturbators, microbes	oxygen	carbon flux	3	3	3
Habitat Maintenance Services							
Landscape linkage & structure/ habitat/refugia	2	deep sea corals, methane seeps	oxygen, temperature, depth, substrate	carbon flux, larval stages	2	2	2
Aesthetic Services							
Spiritual/cultural	2				2	1	0
Aesthetic	2				2	1	3
Recreation	2				2	1	3
Scientific understanding	1	new life forms, microbes, symbioses	depth, sulfide, methane		2	2	2

Deep-Sea Sediments

Deep-sea sediment ecosystems are often ignored when considering the services provided by the ocean. Although human activities continue to expand to greater depths with improved technology, much of the current exploitation (Table 4.2c) is concentrated in the upper 1,000 m. These upper slope sedimentary habitats are repositories for organic carbon moving off the shelf (Walsh et al. 1981) and support expanding commercial and sport fisheries.

Continental slope sediments have higher carbon input and higher abundances of fishes and invertebrates than deeper areas. These are sites of relatively new fisheries for bony fishes such as orange roughy, pelagic armorhead, sablefish, flatfish, and rattails (which occur deeper as well) (Merrett & Haedrich 1997), and for invertebrates such as snow crabs, tanner crabs, golden crabs, northern shrimp, and red crabs (Elner 1982; Otto 1982). Many fisheries have focused on seamounts as well as the continental margin. On seamounts, black and pink corals are harvested for jewelry (Grigg 1993). All of the deepwater fishery taxa are slow-growing, long-lived forms that cannot sustain fishing pressure; most of their populations have declined or will in the near future, and the provisioning of fish secondary production is therefore short-lived and marginal at best (see Snelgrove et al., Chapter 7). Other deep-sea species, such as blue hake, spinetail ray, and spiny eel, have experienced major declines in the past few decades from take as bycatch (i.e., individuals that are removed incidentally as a result of a fishery that is non-selectively targeting some other species) (Baker & Haedrich 2003).

To the extent that biodiversity is considered a valuable resource (e.g., for future uses, scientific interest) in itself, the deep sea functions to maintain and promote high species diversity (Rex 1983; Gage & Tyler 1991). The continental slopes are regions of high diversity, possibly because of the highly heterogeneous environments in space and time. Specific habitats within the deep sea, such as coral (*Lophelia*) reefs (Fossaa et al. 2002), seamounts (Koslow et al. 2001), and some reducing environments (hydrothermal vents, whale falls, and methane seeps) (Van Dover 2000) are recognized as valuable refugia that are important in the maintenance of diversity. More than 99 percent of the deep-sea floor has yet to be sampled (Snelgrove & Smith 2002), so there is considerable potential for future discovery and uses. One emerging area is the exploitation of microbial forms for specific industrial properties, among them their ability to degrade lipids at low temperatures and to break down hydrogen sulfide, and for enzymes to function at high temperatures (Prieur 1997).

Ecological processes that are regulated by deep-sea marine sediment biota include (1) the capture and deposition of organic matter onto the seabed, (2) the transfer of organic matter to higher consumers, (3) the burial of organic matter, and (4) the oxygenation of sediments through bioturbation. In deep-sea sediments, foraminiferans related sarcodines, macrofauna, and nematodes are key bioturbators and regulators of organic cycling. Active suspension and plankton feeders such as sponges, tunicates,

Table 4.2c. The provisioning of goods and services for deep-sea sediment ecosystems.

See Table 4.2a for explanation of ranking scheme.

| | | Biotic | Abiotic | External | Diversity Importance | | |
	Rank	Contributors	Regulators	Interaction	Species	Functional	Habitat
Provisioning Services							
Plants as food	0						
Animals as food	1 to 2	fish, invertebrates, all benthos	oxygen, circulation, substrate	food, life history	3	2	2
Other biological products	0						
Biochemical/medicines/ models for human research	1	microbes, natural products, enzymes	temperature, chemical availability		3	3	3
Fuels/energy	3	microbes	temperature, time		1	1	0
Fiber	1	sponges			1	0	0
C sequestration	1	bioturbators, microbes, infauna	CO_2, temperature, advection	carbon pump	2	3	3
Nonliving materials (geological effects)	0						
Clean seawater	0						

Regulating Services

Service		Components	Processes				
Sediment formation: biodeposition	1	microbes, lithothamnia, biogenic sediments	currents, freshwater and land runoff	sinking of particulates	1	1	0
Nutrient cycling	1	microbes, bioturbators, macrofauna, fishes	circulation, temperature, tides	resuspension	3	3	2
Biological control: disease, invasive species resistance	?				0	0	0
Detoxification, waste disposal	2	microbes, benthos	circulation, resuspension, sedimentation		2	2	1
Climate regulation (C sequestration)	3	bioturbators, microbes, infauna, mobile fauna	hydrodynamic processes, upwelling, resuspension, sedimentation	terrestrial & pelagic input	2	3	3
Food web support processes	1	entire benthos	hydrodynamic processes, upwelling, resuspension, sedimentation, oxygen	terrestrial & pelagic input	3	3	3
Atmosphere composition	1	microbes	oxygen, substrate, turbulent mixing		1	2	2
Flood and erosion control	0						
Redox processes	3	bioturbators, microbes	oxygen	carbon flux	2	3	2

(continued)

Table 4.2c. (continued)

	Rank	Biotic Contributors	Abiotic Regulators	External Interaction	Diversity Importance		
					Species	Functional	Habitat
Habitat Maintenance Services							
Landscape linkage & structure/habitat/refugia	1	deep sea corals, methane seeps	oxygen, temperature, depth, substrate	carbon flux, larval stages	2	2	2
Aesthetic Services							
Spiritual/cultural	1				0	0	0
Aesthetic	1				3	1	2
Recreation	1				0	0	0
Scientific understanding	3	new life forms, microbes, symbioses	depth, sulfide, methane		3	3	3

anemones, and bryozoans capture, ingest, and deposit organic matter or small plankton onto the sea floor in quiescent regions. Passive suspension feeders such as corals, crinoids, selected polychaetes, ophiuroids, and brisingid starfish do the same in higher energy settings. Epibenthic holothurians consume massive deposits of phytodetritus that carpet deep-sea sediments following phytoplankton blooms (Billet 1991), while other surface-deposit feeders are often the first to ingest and transform incoming organic matter into tissue. Nearly all metazoans participate in deep-sea food chains, although diets of most species are unknown (Fauchald & Jumars 1979; Sokolova 2000).

Areas at a depth of greater than 1,000 meters are thought to have reduced biological activity and therefore to be relatively stable compared with shallower ecosystems, and thus they have been a repository for many different kinds of wastes over the last half century (see Snelgrove et al., Chapter 7). However, recent studies show that labile organic matter reaching the deep sea is processed rapidly by benthic macrofauna such as sipunculans and maldanid polychaetes (Graf 1989; Levin et al. 1997), despite low overall faunal biomass (Rowe 1983).

Microbes account for a significant proportion of sediment community oxygen consumption (e.g., 80 percent, Heip et al. 2001), contributing to nutrient cycling through transformation, degradation, and sequestration of organic matter. They control redox conditions within sediments, provide food for protozoan and metazoan consumers (via heterotrophy and symbioses), and their role in nutrient cycling relates strongly to sediment oxygenation (Fenchel & Finlay 1995). Microbes form unusual natural products, enzymes, and detoxification functions (Bunge et al. 2003) that may be exploited commercially. Living microbes have been discovered much deeper in the Earth's crust than any other life form (Parkes et al. 1994).

Key benefits from sediment-based nutrient cycling and carbon burial may include removal of carbon over extended periods of centuries or longer (Heip et al. 2001). The deep sea is currently being considered for more rapid removal of CO_2 in liquid form through direct injection (Ozaki 1997; and see Snelgrove et al., Chapter 7).

Factors Affecting Biodiversity

Numerous abiotic environmental factors influence species diversity (Levin et al. 2001b) and potentially affect processes, goods, and services provided by marine sediments. Salinity, soil texture, organic content, nutrients, waves, currents, and oxygen are abiotic factors that control species composition, densities, and diversity. All of these factors are affected by natural and human-altered regional control of sediment supply, nutrient input, water depth, exposure to disturbance, and hydrologic environment (Diaz & Rosenberg 1995; Parsons et al. 1999; Gray 2002).

Sediment resuspension and motility in shelf and coastal regions is dictated by hydrodynamic processes such as currents, tides, and wave action (Boudreau 1997). This disturbance affects recycling services, the maintenance of sediment oxygenation (e.g.,

Ziebis et al. 1996), and potentially the detoxification of pollutants (Bunge et al. 2003) and rates of biogeochemical cycles (Turner & Millward 2002). It also significantly affects species composition (Ysebaert & Herman 2002).

Hydrodynamic processes primarily determine sediment granulometry and therefore substrate type. This is important to food production, as substrate or habitat availability affects survival of food species of fish and invertebrates (Snelgrove & Butman 1994). Oxygen availability and temperature influence the survival of organisms, reproduction, and function (Garlo 1982), and hence the provision of goods and services by shelf biota. Oxygen availability is particularly important in maintaining sediment redox chemistry (Rhoads et al. 1978; Fenchel & Finlay 1995).

The perception of the deep sea as a species-depauperate and homogeneous habitat has been debunked in the last few decades by evidence of strong regional and temporal variation in the abundance and diversity of deep-sea sediment biota (Levin et al. 2001a; Snelgrove & Smith 2002). The density and biomass of deep-sea infauna are most strongly influenced by organic matter availability (Rowe 1983). Input of organic carbon to the seabed mirrors (but is only a fraction of) surface primary production; it is also influenced strongly by circulation and local flow conditions. Where particulate organic input is high, infaunal species are abundant, animals live deeper in the sediments, and bioturbation rates are greater (Schaff et al. 1992). The continental margins and the north Atlantic are areas of particularly high organic matter input. Topographic features such as seamounts, ridges, canyons, and gullies have accelerated flows where particulate flux is elevated. Because the benthos provides critical trophic support for larger fish and invertebrates, production of harvested species is greatest in these areas, as are rates of carbon processing, burial, and sequestration.

In some estuarine and shelf areas, excess production from surface waters can lead to hypoxia in bottom waters (see discussion of nutrient loading in Chapter 7). An intriguing parallel occurs in some deep-sea areas when high production from surface waters sinks to bottom areas with sluggish circulation, leading to the formation of midwater oxygen minimum zones (OMZs) at depths of 100 to 1,000 meters. Within OMZs, there is reduced productivity, less remineralization of carbon, and lowered functional and species diversity of the sediment biota. These effects occur over huge areas ($>10^6$ km^2) of the sea floor (Levin 2003). Temporal changes in the boundaries of OMZs exert tremendous control on seabed productivity and diversity over ecological time (e.g., with El Niño events; Arntz et al. 1988) and over geological time (Rogers 2000).

The structure and function of deep-sea sediment biota is also influenced by benthic storms (Hollister & McCave 1984) and turbidity flows or mass wasting (Masson et al. 1996). Microbial function and activity are greatly influenced by availability of oxygen, organic matter, and reduced compounds such as methane and sulfide. Amazing discoveries of microbial syntrophy (symbioses involving microbes of different metabolic func-

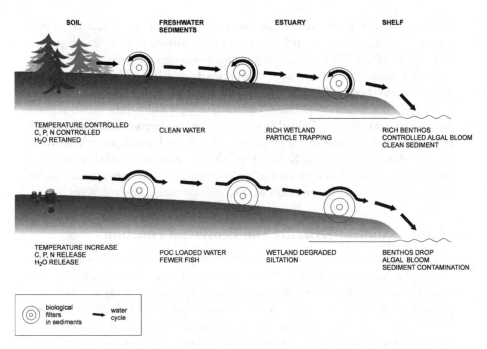

Figure 4.1. Schematic depiction of interrelated nature of soil, freshwater, and coastal marine sedimentary ecosystems. The top diagram depicts a functioning ecosystem prior to deforestation. The lower diagram illustrates the cascade of changes that may occur from disturbance to soils. Deep-sea ecosystems are not shown because their linkages with terrestrial and freshwater domains are indirect and expressed only at long temporal and large spatial scales. Arrows indicate flow of materials (water, nutrients, organic matter), and circles indicate biological filters. POC is particulate organic carbon, C is carbon, P is phosphorus, and N is nitrogen.

tions), multiple bacterial symbioses within invertebrates, and sediment ecosystems reliant on methane for carbon have come from highly reduced sediments in the deep sea.

Linkages to Marine Sedimentary Systems

Marine sedimentary goods and services are linked to adjacent ecosystems, including the water column above, the coastal zone, and even freshwater systems (Figure 4.1; see also Chapter 1, Figure 1.1). Because they are open transitional systems between land and sea, estuaries and their associated biotic components have direct hydrological links to coastal seas and upland watersheds. Tides provide the principal natural vector for marine-derived inputs to estuaries and freshwater flows from surface or groundwater sources that convey materials, nutrients, and organisms from upland drainage basins.

Sedimentary inputs to estuaries may be from either marine or upland sources, whereas biological linkages occur through movement of organisms in and out of estuaries (Levin et al. 2001a). The food provision and food web supporting services of estuarine and shelf sediments are closely linked with the overlying pelagic realm and particularly their food webs (Steele 1974). Many benthic invertebrates and fishes spend the first part of their life cycle within the plankton, providing linkages with pelagic species through predator-prey interactions (Bullard et al. 1999). Among deep-sea taxa, this is true of many commercially harvested taxa such as snow crab, golden crab, armorhead, sablefish, and grenadiers (Zheng & Kruse 2000). The terrestrial linkage of supplying estuaries and coasts with input of detritus and nutrients are also important to trophic support processes and food provision from estuarine through slope sediments.

Marine Sediment Diversity and Ecosystem Function

The role of species diversity in regulating ecosystem processes and services in sedimentary systems has received considerably less attention than its role in terrestrial systems (Estes & Peterson 2000). Although there are many examples of living organisms that play critical roles in providing services and functions, there is little evidence that biodiversity *per se* is critical for the delivery of services and functions. In many instances, it is likely that the availability of specific functional groups is most important in providing a given service or function (Tables 4.2a–4.2c). The benthic biota of estuaries are the least diverse of the marine sediment realms, but specific groups perform valuable functions: they create habitat, trap and retain sediments (e.g., rooted vegetation), maintain water quality (e.g., filter-feeding bivalves), contribute to aeration of subsurface sediments (e.g., bioturbators/burrowing crabs), and shunt production from the microbial decomposers to higher trophic levels (e.g., grazing snails and amphipods). There is some evidence that diversity decreases variability in rates of nutrient recycling and there are complementary effects of diversity on function, but there is no consistent relationship between species richness and function (Emmerson et al. 2001; see also Biles et al. 2003). However, few experiments to test these questions have been conducted in marine systems. Experiments with hard substrate communities have suggested that species diversity enhances resistance to invasive species (Stachowicz et al. 1999), but similar experiments are lacking for estuarine sediments. Nonetheless, estuaries have a public visibility that seems to confer a high value to the limited species diversity for aesthetic, recreational, and scientific reasons.

The role of species diversity on the continental slope is not well documented in provision of trophic support, nutrient cycling, and waste disposal/detoxification, but it is clear that multiple species are involved. In cases where multiple species are eliminated by hypoxic events, for example, the loss of key sedimentary functions has resulted (Rabalais et al. 1996), but it could be argued that loss of functional groups, rather than species, is more important (Elmgren & Hill 1997). In estuarine and shelf ecosystems, the diversity of structure-forming species often contributes to habitat diversity, which subsequently increases the diversity of species that utilize that habitat and therefore may enhance key services such as food production (Auster et al. 1996).

The high diversity of infaunal species in the deep sea raises many questions about rates and redundancy that are largely unanswered (Snelgrove & Smith 2002). The relative importance of species diversity for the efficiency of the deep-sea functions discussed above has not been tested experimentally. In general, measures of macrofaunal density, biomass, or diversity have been poor predictors of functions such as bioturbation, whereas particulate organic carbon (POC) flux and densities of selected megafauna can be good predictors (Smith 1992; Smith & Rabouille 2002).

One formidable challenge is to determine whether diversity at the level of habitats, functional groups, species, genes, or gene expression (functional genomics) is most critical for sustaining ecological processes and services. Recent research has considered the role of landscape configuration (Archambault & Bourget 1999) and the effect of anthropogenic modifications and structures on estuarine biodiversity (Chapman & Bulleri 2003). These foci have potential applications for restoring and conserving biodiversity in the face of growing pressures for increased coastal development; they also have potential consequences for processes and services.

Theory based on the terrestrial literature suggests that if each species performs a function slightly differently, then sediments with high diversity are likely to achieve the most effective function (i.e., sampling effect) (Loreau et al. 2001; Zedler et al. 2001). Interspecific facilitative interactions are particularly likely to enhance functions in areas with low oxygen, high sulfides, food scarcity, physical disturbance, or other stressors (Levin et al. 2001b). Structures on the sea floor such as polychaete feeding mounds, tracks in sediments from surface burrowers, and discarded shell material provide heterogeneity, which facilitates adults and juveniles of many deep-sea species, providing food, substrate, and refugia (Levin et al. 1997; Snelgrove & Smith 2002).

Research Needs and Recommendations

The vast majority of marine sedimentary organisms are undescribed and unknown (e.g., 10 million macrofaunal species are estimated in Grassle & Maciolek 1992), with the diversity of the smaller organisms much less well understood than that of larger organisms. There is a fundamental need to document the taxonomic composition of sedimentary biota through biodiversity surveys of representative marine habitats. Although the large area of marine sedimentary habitat precludes a comprehensive biodiversity survey, it is reasonable to survey representative areas in order to generate diversity estimates for different habitat types and biogeographic maps for relatively common species. This information is critical to manage and conserve the functional properties of marine ecosystems for the long term, particularly in areas that are vulnerable to human activities (see Snelgrove et al., Chapter 7; Wall et al. 2001). A significant obstacle to the study of biodiversity is the "taxonomic impediment"—a worldwide shortage of taxonomists (Hoagland 1995; Environment Australia 1998).

The role of marine sediment biodiversity in the regulation of ecosystem processes and services is poorly understood, particularly for groups such as the fungi, protists, and meiofauna. Even for macrofauna and megafauna, the role of biodiversity has been examined in only a few studies. Levels of functional redundancy within and across groups and their relative importance must be characterized to offer predictive capabilities concerning controls on, and threats to, ecosystem processes. Given the many abiotic variables that influence biodiversity patterns and the linkages between different sedimentary ecosystems, studies of ecosystem processes and services must consider marine sediments and their biodiversity when establishing and implementing marine conservation strategies. Finally, efforts to value sedimentary biota are effectively nonexistent. Lack of direct experience alone limits our capacity to value marine sedimentary services. Aside from coral reefs, sandy beaches, and wetlands, most sedimentary habitats generate little public concern and hence often rate low in conservation priority. This situation can be altered as both scientists and the public improve their understanding of the critical roles and services provided by marine sediments in the biosphere.

Literature Cited

Archambault, P., and E. Bourget. 1999. Influence of shoreline configuration on spatial variation of meroplanktonic larvae, recruitment and diversity of benthic subtidal communities. *Journal of Experimental Marine Biology and Ecology* 238:161–184.

Arntz, W.E., E. Valdivia, and J. Zeballos. 1988. Impact of El Nino 1982–83 on the commercially exploited invertebrates (mariscos) of the Peruvian shore. *Meeresforschung* 32:3–22.

Auster, P.J., R.J. Malatesta, R.W. Langton, L. Watling, P.C. Valentine, C.L.S. Donaldson, E.W. Langton, A.N. Shepard, and I.G. Babb. 1996. The impacts of mobile fishing gear on seafloor habitats in the Gulf of Maine (northwest Atlantic): Implications for conservation of fish populations. *Reviews in Fisheries Science* 4:185–202.

Baker, K., and R.L. Haedrich. 2003. Could some deep-sea fishes be species-at-risk? International Deep-Sea Biology Conference, Aug. 2003, Coos Bay, Oregon. (Abstract)

Beach, D. 2002. *Coastal Sprawl: The Effects of Urban Design on Aquatic Ecosystems in the United States.* Arlington, Virginia, Pew Oceans Commission. http://www.pewoceans.org/oceanfacts/2002/04/12/fact_25649.asp.

Biles, C.L., M. Solan, I. Isaksson, D.M. Paterson, C. Emes, and D.G. Raffaelli. 2003. Flow modifies the effect of biodiversity on ecosystem functioning: An *in situ* study of estuarine sediments. *Journal of Experimental Marine Biology and Ecology* 285–286:165–177.

Billett, D.S.M. 1991. Deep-sea holothurians. *Oceanography and Marine Biology: An Annual Review* 29:259–317.

Boudreau, B.P. 1997. A one-dimensional model for bed-boundary layer particle exchange. *Journal of Marine Systems* 11:279–303.

Boyd, T.J., and A.F. Carlucci. 1996. Rapid microbial degradation of phenolic materials in California (USA) coastal environments. *Aquatic Microbial Ecology* 11:171–179.

Bullard, S.G., N.L. Lindquist, and M.E. Hay. 1999. Susceptibility of invertebrate larvae to predators: How common are post-capture larval defenses? *Marine Ecology Progress Series* 191:153–161.

Bunge, M., L. Adrian, A. Kraus, M. Opel, W.G. Lorenz, J.R. Andreesen, H. Goerisch, and U. Lechner. 2003. Reductive dehalogenation of chlorinated dioxins by an anaerobic bacterium. *Nature* 421:357–360.

Burke, L., Y. Kura, K. Kassem, C. Revenga, M. Spalding, and D. McAllister. 2001. *Pilot Analysis of Global Ecosystems: Coastal Ecosystems.* Washington, DC, World Resources Institute.

Carlson, J., T. Randall, and M. Mroczka. 1997. Feeding habits of winter flounder (*Pleuronectes americanus*) in a habitat exposed to anthropogenic disturbance. *Journal of Northwest Atlantic Fishery Science* 21:65–73.

Chapman, M.G., and F. Bulleri. 2003. Intertidal seawalls: New features of landscape in intertidal environments. *Landscape and Urban Planning* 62:159–172.

Costanza, R. 1999. The ecological, economic and social importance of the oceans. *Ecological Economics* 31:199–213.

Costanza, R., R. d'Arge, R. de Groot, S. Farber, M. Grasso, B. Hannon, K. Limburg, S. Naeem, R.V. O'Neill, J. Paruelo, R.G. Raskin, P. Sutton, and M. van den Belt. 1997. The value of the world's ecosystem services and natural capital. *Nature* 387:253–260.

Covi, M.P., and R.T. Kneib. 1995. Intertidal distribution, population dynamics and production of the amphipod *Uhlorchestia spartinophila* in a Georgia, USA, salt marsh. *Marine Biology* 121:447–455.

Daily, G.C., T. Söderqvist, S. Aniyar, K. Arrow, P. Dasgupta, P.R. Ehrlich, C. Folke, A. Jansson, B.-O. Jansson, N. Kautsky, S. Levin, J. Lubchenco, K.-G. Mäler, D. Simpson, D. Starrett, D. Tilman, and B. Walker. 2000. The value of nature and the nature of value. *Science* 289:395–396.

Dasgupta, P., S. Levin, and J. Lubchenco. 2000. Economic pathways to ecological sustainability. *BioScience* 50:340–345.

Diaz, R.J., and R. Rosenberg. 1995. Marine benthic hypoxia: A review of its ecological effects and the behavioural responses of benthic macrofauna. *Oceanography and Marine Biology: An Annual Review* 33:245–303.

Elmgren, R., and C. Hill. 1997. Ecosystem function at low biodiversity: The Baltic example. In: *Marine Biodiversity: Patterns and Processes,* edited by R.F.G. Ormond, J.D. Gage, and M.V. Angel, pp. 319–335. Cambridge, UK, Cambridge University Press.

Elner, R.W. 1982. Overview of the snow crab *Chionoecetes opilio* fishery in Atlantic Canada. *Proceedings of the International Symposium on the Genus Chionoecetes,* May 3–6, 1982, pp. 3–19. Lowell Wakefield Symposia Series, Alaska Sea Grant Report. Alaska Sea Grant Program.

Emmerson, M.C., M. Solan, C. Emes, D.M. Paterson, and D. Raffaelli. 2001. Consistent patterns and the idiosyncratic effects of biodiversity in marine ecosystems. *Nature* 411:73–77.

Environment Australia. 1998. *The Darwin Declaration.* Australian Biological Resources Study. Canberra, Environment Australia.

Estes, J.A., and C.H. Peterson. 2000. Marine ecological research in seashore and seafloor systems: Accomplishments and future directions. *Marine Ecology Progress Series* 195:281–289.

Ewel K.C., C. Cressa, R.T. Kneib, P.S. Lake, L.A. Levin, M.A. Palmer, P. Snelgrove, and D.H. Wall. 2001. Managing critical transition zones. *Ecosystems* 4:452–460.

Fauchald, K., and P.A. Jumars. 1979. The diet of worms: A study of polychaete feeding guilds. *Oceanography and Marine Biology* 17:193–284.

Feder, H.M., and T.H. Pearson. 1988. The benthic ecology of Loch Linnhe and Loch Eil, a sea-loch system on the west coast of Scotland. V. Biology of the dominant soft-bottom epifauna and their interaction with the infauna. *Journal of Experimental Marine Biology and Ecology* 116:99–134.

Fenchel, T., and B.J. Finlay. 1995. *Ecology and Evolution in Anoxic Worlds.* Oxford Series in Ecology and Evolution. Oxford, UK, Oxford University Press.

Fonseca, M.S., and J.S. Fisher. 1986. A comparison of canopy friction and sediment movement between four species of seagrass with reference to their ecology and restoration. *Marine Ecology Progress Series* 29:15–22.

Fossaa, J., P. Mortensen, and D. Furevik. 2002. The deep-water coral *Lophelia pertusa* in Norwegian waters: Distribution and fishery impacts. *Hydrobiologia* 471:1–12.

Gage, J.D., and P.A. Tyler. 1991. *Deep-Sea Biology: A Natural History of Organisms at the Deep-Sea Floor.* Cambridge, UK, Cambridge University Press.

Gallagher, E.D., and K.E. Keay. 1998. Organism-sediment-contaminant interactions in Boston Harbor. In: *Contaminated Sediments in Boston Harbor,* edited by K.D. Stolzenbach and E.E. Adams, pp. 89–132. Cambridge, Massachusetts, MIT Sea Grant College Program.

Garlo, E.V. 1982. Increase in a surf clam population after hypoxic water conditions off Little Egg Inlet, New Jersey. *Journal of Shellfish Research* 2:59–64.

Geiselbrecht, A.D., R.P. Herwig, J.W. Deming, and J.T. Staley. 1996. Enumeration and polyphyletic analysis of polycyclic aromatic hydrocarbon-degrading marine bacteria from Puget Sound sediments. *Applied Environmental Microbiology* 62:3344–3349.

Graf, G. 1989. Pelagic-benthic coupling in a deep sea benthic community. *Nature* 341:437–439.

Grant, J., and G. Gust. 1987. Prediction of coastal sediment stability from photopigment content of mats of purple sulphur bacteria. *Nature* 330:244–246.

Grant, W.D., L.F. Boyer, and L.P. Sanford. 1982. The effects of bioturbation on the initiation of motion of intertidal sands. *Journal of Marine Research* 40:659–677.

Grassle, J.F., and N.J. Maciolek. 1992. Deep-sea species richness: Regional and local diversity estimates from quantitative bottom samples. *American Naturalist* 139:313–341.

Gray, J.S. 2002. Species richness of marine soft sediments. *Marine Ecology Progress Series* 244:285–297.

Grigg, R.W. 1993. Precious coral fisheries of Hawaii and the U.S. Pacific islands. *Marine Fisheries Review* 55:50–60.

Hanson, R.B., H.W. Ducklow, and G.G. Field, editors. 2000. *International Geosphere-Biosphere Programme Book Series,* No. 5. Cambridge, UK, Cambridge University Press.

Heip, C.H.R., G. Duineveld, E. Flach, G. Graf, W. Helder, P.M.J. Herman, M. Lavaleye, J.J. Middelburg, O. Pfannkuche, K. Soetaert, K. Soltwedel, H. de Stigter, L. Thomsen, J. Vanaverbeke, and P. de Wilde. 2001. The role of the benthic biota in sedimentary metabolism and sediment-water exchange processes in the Goban Spur area (NE Atlantic). *Deep-Sea Research* Part II 48:3223–3243.

Henriksen, K., M.B. Rasmussen, and A. Jensen. 1983. Effect of bioturbation on microbial nitrogen transformations in the sediment and fluxes of ammonium and nitrate for the overlying water. *Ecological Bulletin* 35:193–205.

Heymans, J.J., and D. Baird. 1995. Energy flow in the Kromme Estuarine ecosystem, St. Francis Bay, South Africa. *Estuarine, Coastal and Shelf Science* 41:39–59.

Hoagland, K.E. 1995. *The Taxonomic Impediment and the Convention on Biodiversity.* http://www.science.uts.edu.au/sasb/TaxImp.html.

Hollister, C.D., and I.N. McCave. 1984. Sedimentation under deep-sea storms. *Nature* 309:220–225.

Koslow, J.A., K. Gowlett-Holmes, J.K. Lowry, T. O'Hara, G.C.B. Poore, and A. Williams. 2001. Seamount benthic macrofauna off southern Tasmania: Community structure and impact of trawling. *Marine Ecology Progress Series* 213:111–125.

Kruczynski, W. 1999. The importance of coastal wetlands: Why do we need to protect them? *ASB Bulletin* 46:246–272.

Lambshead, P.J.D. 1993. Recent developments in marine benthic biodiversity research. *Oceanis* 19:5–24.

Laurel, B.J., R.S. Gregory, and J.A. Brown. 2003. Predator distribution and habitat patch area determine predation rates on Age-0 juvenile cod *Gadus* spp. *Marine Ecology Progress Series* 251:245–254.

Lee, R.F., and D.S. Page. 1997. Petroleum hydrocarbons and their effects in subtidal regions after major oil spills. *Marine Pollution Bulletin* 34:928–940.

Lee, R.W., D.W. Kraus, and J.E. Doeller. 1999. Oxidation of sulfide by *Spartina alterniflora* roots. *Limnology and Oceanography* 44:1155–1159.

Levin, L.A. 2003. Oxygen minimum zone benthos: Adaptation and community response to hypoxia. *Oceanography and Marine Biology: An Annual Review* 41:1–45.

Levin, L.A., N. Blair, D.J. DeMaster, G. Plaia, W. Fornes, C. Martin, and C. Thomas. 1997. Rapid subduction of organic matter by maldanid polychaetes on the North Carolina slope. *Journal of Marine Research* 55:595–611.

Levin, L.A., D.F. Boesch, A. Covich, C. Dahm, C. Erseus, K.C. Ewel, R.T. Kneib, A. Moldenke, M.A. Palmer, P. Snelgrove, D. Strayer, and J.M. Weslawski. 2001a. The function of marine critical transition zones and the importance of sediment biodiversity. *Ecosystems* 4:430–451.

Levin, L.A., R.J. Etter, M.A. Rex, A.J. Gooday, C.R. Smith, J. Pineda, C.T. Stuart, R.R. Hessler, and D. Pawson. 2001b. Environmental influences on regional deep-sea species diversity. *Annual Review of Ecology and Systematics* 132:51–93.

Levinton, J.S. 1995. Bioturbators as ecosystem engineers: Population dynamics and material fluxes. In: *Linking Species and Ecosystems*, edited by C.G. Jones and J.H. Lawton, pp. 29–36. New York, Chapman and Hall.

Lindley, J.A., J.C. Gamble, and H.G. Hunt. 1995. A change in the zooplankton of the central North Sea (55 degrees to 58 degrees N): A possible consequence of changes in the benthos. *Marine Ecology Progress Series* 119:299–303.

Loreau, M., S. Naeem, P. Inchausti, J. Bengtsson, J.P. Grime, A. Hector, D.U. Hooper, M.A. Huston, D. Raffaelli, B. Schmid, D. Tilman, and D.A. Wardle. 2001. Biodiversity and ecosystem functioning: Current knowledge and future challenges. *Science* 294:804–808.

Martin, J.H., K.H. Coale, K.S. Johnson, S.E. Fitzwater, R.M. Gordon, S.J. Tanner, C.N. Hunter, V.A. Elrod, J.L. Nowicki, T.L. Coley, R.T. Barber, S. Lindley, A.J. Watson, K. Vanscoy, C.S. Law, M.I. Liddicoat, R. Ling, T. Stanton, J. Stockel, C. Collins, A. Anderson, R. Bidigare, M. Ondrusek, M. Latasa, F.J. Millero, K. Lee, W. Yao, J.Z. Zhang, G. Friederich, C. Sakamoto, F. Chavez, K. Buck, Z. Kolber, R. Greene, P.

Falkowski, S.W. Chisholm, F. Hoge, R. Swift, J. Yungel, S. Turner, P. Nightingale, A. Hatton, P. Liss, and N.W. Tindale. 1994. Testing the iron hypothesis in ecosystems of the equatorial Pacific Ocean. *Nature* 371:123–129.

Masson, D.G., N.H. Kenyon, and P.P.E. Weaver. 1996. Slides, debris flows and turbidity currents. In: *Oceanography: An Illustrated Guide,* edited by C.P. Summerhayes and S.A. Thorpe, pp. 136–151. London, Manson.

Merrett, N.R., and R.L. Haedrich. 1997. *Deep-Sea Demersal Fish and Fisheries.* London, Chapman and Hall.

Otto, R.S. 1982. An overview of the eastern Bering Sea tanner crab fisheries. *Proceedings of the International Symposium on the Genus Chionoecetes,* May 3–6, 1982. Lowell Wakefield Symposia Series, 1982, pp. 83–115, Alaska Sea Grant Report. Alaska Sea Grant Program.

Ozaki, M. 1997. CO_2 injection and dispersion in mid-ocean depth by moving shop. *Waste Management* 17:369–373.

Parkes, R.J., B.A. Cragg, S.J. Bale, J.M. Getliff, K. Goodman, P.A. Rochelle, J.C. Fry, A.J. Weightman, and S.M. Harvey. 1994. Deep bacterial biosphere in Pacific Ocean sediments. *Nature* 37:410–413.

Parsons, M.L., Q. Dortch, R.E. Turner, and N.N. Rabalais. 1999. Salinity history of coastal marshes reconstructed from diatom remains. *Estuaries* 22:1078–1089.

Paterson, D.M., and K.S. Black. 1999. Water flow, sediment dynamics and benthic biology. *Advances in Ecological Research* 29:155–193.

Paterson, G.L.J. 1993. Patterns of polychaete assemblage structure from bathymetric transects in the Rockall Trough, NE Atlantic Ocean. Ph.D. thesis. University of Wales. 252 pp.

Pelegri, S.P., and T.H. Blackburn. 1995. Effect of bioturbation by *Nereis* sp., *Mya arenaria* and *Cerastoderma* sp. on nitrification and denitrification in estuarine sediments. *Ophelia* 42:289–299.

Pournelle, J.R. 2003. The littoral foundations of the Uruk state: Using satellite photography toward a new understanding of the 5th/4th millennium BCE landscapes in the Warka Survey Area, Iraq. In: *Chalcolithic and Early Bronze Age Hydrostrategies,* edited by D. Gheorghiu, pp. 5–23. BAR International Series 1123. Oxford, UK, Archaeopress.

Prieur, D. 1997. Microbiology of deep-sea hydrothermal vents. *Trends in Biotechnology* 15:242–244.

Rabalais, N.N., W.J. Wiseman, Jr., R.E. Turner, D. Justic, B.K. Sen Gupta, and Q. Dortch. 1996. Nutrient changes in the Mississippi River and system responses on the adjacent continental shelf. *Estuaries* 19:386–407.

Rex, M.A. 1983. Geographic patterns of diversity in deep-sea benthos. In: *The Sea,* edited by G.T. Rowe, Vol. 8. pp. 453–472. New York, Wiley Interscience.

Rhoads, D.C. 1963. Rates of sediment reworking by *Yoldia limatula* in Buzzard's Bay, Massachusetts, and Long Island Sound. *Journal of Sedimentary Petrology* 33:723–727.

Rhoads, D.C., P.L. McCall, and J.Y. Yingst. 1978. Disturbance and production on the estuarine seafloor. *American Scientist* 66:577–586.

Rhoads, D.C., and D.K. Young. 1970. The influence of deposit-feeding organisms on sediment stability and community trophic structure. *Journal of Marine Research* 28:150–178.

Rogers, A.D. 2000. The role of the oceanic oxygen minima in generating biodiversity in the deep sea. *Deep-Sea Research* Part II 47:119–148.

Rowe, G.T. 1983. Biomass and production of the deep-sea macrobenthos. In: *The Sea,* edited by G.T. Rowe, Vol. 8 pp. 97–121. New York, Wiley Interscience.

Schaff, T., L. Levin, N. Blair, D. DeMaster, R. Pope, and S. Boehme. 1992. Spatial heterogeneity of benthos on the Carolina continental slope: Large (100 km)-scale variation. *Marine Ecology Progress Series* 88:143–160.

Smith, C.R. 1992. Factors controlling bioturbation in deep-sea sediments and their relation to models of carbon diagenesis. In: *Deep Sea Food Chains and the Global Carbon Cycle,* NATO Asi Science Series C: Mathematics and Physical Science, Vol. 360, pp. 375–393. Dordrecht, The Netherlands, Kluwer Academic Publishers.

Smith, C.R., and C. Rabouille. 2002. What controls the mixed layer depth in deep-sea sediments? The importance of POC Flux. *Limnology and Oceanography* 47:418–426.

Snelgrove, P.V.R. 1999. Getting to the bottom of marine biodiversity: Sedimentary habitats. *BioScience* 49:129–138.

Snelgrove, P.V.R., T.H. Blackburn, P. Hutchings, D. Alongi, J.F. Grassle, H. Hummel, G. King, I. Koike, P.J.D. Lambshead, N.B. Ramsing, and V. Solis-Weiss. 1997. The importance of marine biodiversity in ecosystem processes. *Ambio* 26:578–583.

Snelgrove, P.V.R., and C.A. Butman. 1994. Animal-sediment relationships revisited: Cause versus effect. *Oceanography and Marine Biology: An Annual Review* 32:111–177.

Snelgrove, P.V.R., and C.R. Smith. 2002. A riot of species in an environmental calm: The paradox of the species-rich deep-sea. *Oceanography and Marine Biology: An Annual Review* 40:311–342.

Sokolova, M.N. 2000. *Feeding and Trophic Structure of the Deep-Sea Macrobenthos.* Enfield, New Hampshire, Science Publisher Inc.

Srinivasan, M., and B.A. Mahajan. 1989. Mercury pollution in an estuarine region and its effect on a coastal population. *International Journal of Environmental Studies* A & B 35:63–69.

Stachowicz, J.J., R.B. Whitlatch, and R.W. Osman. 1999. Species diversity and invasion resistance in a marine ecosystem. *Science* 286:1577–1579.

Steele, J.H. 1974. *The Structure of Marine Ecosystems.* Oxford, UK, Blackwell Scientific.

Thayer, G.W., D.R. Colby, and W.F. Hettler, Jr. 1987. Utilization of the red mangrove prop root habitat by fishes in South Florida. *Marine Ecology Progress Series* 35:25–38.

Turner, A., and G.E. Millward. 2002. Suspended particles: Their role in estuarine biogeochemical cycles. *Estuarine, Coastal and Shelf Science* 55:857–883.

United Nations. 1992. *United Nations Framework Convention on Global Climate Change.* New York, United Nations.

Van Dover, C.L. 2000. *The Ecology of Hydrothermal Vents.* Princeton, New Jersey, Princeton University Press.

Wall, D.H., P.V.R. Snelgrove, and A.P. Covich. 2001. Conservation priorities for soil and sediment invertebrates. In: *Conservation Biology,* edited by M.E. Soulé and G.H. Orians, pp. 99–123. Washington, DC, Island Press.

Walsh, J.W., G.T. Rowe, C.P. McRoy, and R. Iverson. 1981. Biological export of shelf carbon is a sink of the global CO_2 cycle. *Nature* 291:196–201.

Weslawski, J.M., B. Urban-Malinga, L. Kotwicki, K.W. Opalinski, M. Szymelfenig, and M. Dutkowski. 2000. Sandy coastlines: Are there conflicts between recreation and natural values? *Oceanological Studies* 29:5–18.

Wright, L.D., L.C. Schaffner, and J.P.Y. Maa. 1997. Biological mediation of bottom boundary layer processes and sediment suspension in the lower Chesapeake Bay. *Marine Geology* 141:27–50.

Ysebaert, T., and P.M.J. Herman. 2002. Spatial and temporal variation in benthic macrofauna and relationships with environmental variables in an estuarine, intertidal soft sediment environment. *Marine Ecology Progress Series* 244:105–124.

Zedler, J.B., J.C. Callaway, and G. Sullivan. 2001. Declining biodiversity: Why species matter and how their functions might be restored in Californian tidal marshes. *BioScience* 51:1005–1017.

Zheng, J., and K.H. Kruse. 2000. Recruitment patterns of Alaskan crabs in relation to decadal shifts in climate and physical oceanography. *ICES Journal of Marine Science* 57:438–451.

Ziebis, W., M. Huettel, and S. Forster. 1996. Impact of biogenic sediment topography on oxygen fluxes in permeable seabeds. *Marine Ecology Progress Series* 140:227–237.

PART II

Assessment of the Vulnerability of Critical Below-Surface Habitats, Functions, and Taxa

Paul V. R. Snelgrove

In the previous chapters, we described the many goods and services provided by soil and sediment biota and alluded briefly to some of the effects that human activities have had or will likely have on the continued delivery of these services. The next section focuses specifically on the *vulnerability* of ecosystem goods and services, defined here as the probability that ecosystem services will be altered by external disturbances. Each of the chapters that follow notes that different threats operate on different spatial and temporal scales, as does the provisioning of different ecosystem goods and services. For example, invasive species are identified as major stressors in terrestrial, freshwater, and marine ecosystems, but some invasive species are localized in their current distribution and may disperse rapidly or slowly. By contrast, global climate change is large scale by definition, and it cuts across habitats, ecosystems, and even the terrestrial/freshwater/marine interface. Decomposition, a key regulation service in all of the domains, occurs over very broad landscapes, aquatic habitats, and seascapes, but it is manifested largely at the scale of individual microbes and invertebrates. Differences in exposure (frequency, intensity), sensitivity to exposure (likelihood of change), and resilience (capacity to rebound from alteration) are all factors that are scale-dependent and vary among ecosystems, habitats, and the specific organisms that deliver goods and services.

The initial goal of these chapters was to address three broad questions on vulnerability of ecosystem goods and services. First, what is the vulnerability of critical below-surface ecosystems, habitats, functions, and taxa to human activities? Second, to what extent is the vulnerability of different systems spatial- and time-scale dependent? Third, what are the implications for management when we consider the vulnerability of soils and sediments, their biodiversity, and their provisioning of ecosystem goods and services?

The evidence for linkages between vulnerability of goods and services and biodiversity varies considerably among ecosystems. However, in the chapters that follow, we provide the most relevant examples and case studies currently available in order to illustrate the issues and to point out where information is inadequate. Ultimately the initial questions we posed cannot be fully addressed with our current knowledge, but the situation is rapidly improving. Questions regarding the biotic basis of ecosystem processes and the implications of biodiversity loss and associated effects on delivery of goods and services are now at the forefront of conservation biology. Unfortunately, the magnitude and pervasiveness of many threats stemming from human activities is rapidly increasing at a rate that underscores the urgency of research in this area.

5

Vulnerability to Global Change of Ecosystem Goods and Services Driven by Soil Biota

David A. Wardle, Valerie K. Brown, Valerie Behan-Pelletier, Mark St. John, Todd Wojtowicz, Richard D. Bardgett, George G. Brown, Phillip Ineson, Patrick Lavelle, Wim H. van der Putten, Jonathan M. Anderson, Lijbert Brussaard, H. William Hunt, Eldor A. Paul, and Diana H. Wall

Soil biota play an essential role in the delivery of a range of ecosystem goods and services. However, it is important to recognize that ecosystems are not static, and that human-induced global change phenomena have the potential to alter the capacity of soil organisms to provide this contribution. Given that global change phenomena directly or indirectly impact the soil biota in some way (Wolters et al. 2000; Wardle 2002), the question that emerges is how these phenomena may alter the goods and services, driven by soil organisms, upon which we all depend.

Predicting the effects of global change on ecosystem goods and services requires explicit acknowledgment of the vulnerability of ecosystems, and therefore of the organisms that drive those ecosystems, to global change. This chapter focuses on the vulnerability of goods and services to global change. In assessing this, we use the definition of *vulnerability* provided by the Resilience Alliance (www.resalliance.org), which is "the propensity of social and ecological systems to suffer harm from exposure to external stresses and shocks," a definition that involves three components: (1) exposure to events and stresses, (2) sensitivity to such exposures, and (3) resilience owing to adaptive measures to anticipate and reduce future harm. We first discuss conceptual issues regarding vulnerability of soil organisms and processes to global change in terms of effects of spatial scale, the extrinsic and intrinsic determinants of vulnerability, and the mechanistic bases of how the belowground subsystem responds to global change. We then demonstrate, through worked examples, the degree of vulnerability of those ecosystem goods and services driven by soil biota to selected agents of global change.

The Overarching Role of Spatial and Temporal Scale

The study of the vulnerability of ecosystems to global change requires explicit consideration of spatial and temporal scale. Agents of global change simultaneously operate over a range of scales; some operate mainly in the short term at local scales while others are more pervasive (Table 5.1). Further, ecosystem services are delivered at different scales of time and space from the immediate release of nutrients in the vicinity of a root tip to infiltration of water and storage in capillary porosity for periods of months to years.

Ecological processes that sustain provision of these services also operate at a variety of scales. The diversity of scales and the match between scaling of processes and their outputs may significantly influence their vulnerability.

Soil ecosystem services depend either on the direct and immediate outputs of organism activities (e.g., nutrient release from digestive processes) or their longer-term effects on soil physical properties (e.g., water retention in soil pores built by bioturbators). They may also depend on the direct effect of abiotic processes that operate at large scales (e.g., porosity created by alternations of drying and rewetting cycles). Biological processes that sustain ecosystem services may operate at four different scales of time and space depending on their nature and location (Lavelle 1997):

1. *Short-term digestion-associated processes.* Digestion occurs in the immediate vicinity of microorganisms where exoenzymes are active, in the guts of invertebrates or in the rhizosphere soil close to active root tips. Such microsites are a few cubic microns to millimeters in volume and processes develop during periods of hours to a few days.
2. *Intermediate phase in fresh biogenic structures.* Microbial activation triggered during gut transit or mechanical mixing of organic materials with soil culminates in fresh biogenic structures, such as fresh earthworm casts or termite fecal pellets. Activity then progressively decreases during the few days or weeks following deposition.
3. *Longer-term scale of stabilized biogenic structures.* Some structures created by invertebrates or roots are highly compact. These structures are the components of stable macroaggregate structures that determine soil hydraulic properties and resistance to erosion (Blanchart et al. 1999; Chauvel et al. 1999). Their life span may extend over periods of months to years depending on their composition and the dynamics of soil structural features (Decaëns 2000).
4. *Soil profiles.* Biogenic structures combine with other structures and elements of soil to form soil horizons. In some cases, creeping of soil along slopes may be triggered

Table 5.1. Vulnerability matrix, showing how different global change drivers have effects on the belowground subsystem that may be manifested at different spatial scales.

Note that some drivers operate mainly at local spatial scales while others are more pervasive.

Scale	*Global Change Driver*					
	Land Use	Agricultural Intensification	Climate Change	Pollution	Invasive Organisms	Urbanization
Global	X		X			
Regional	X	X	X		X	
Landscape	X	X	X		X	X
Field	X	X	X	X	X	X
Patch	X	X	X	X	X	X
Single Unit	X	X	X	X	X	X

by an accumulation of surface deposits by soil invertebrates (Nye 1955). As long-term plant successional processes and pedogenesis operate, changes occur in soil organism communities and their effects on soil structure over whole watersheds and timescales of years to centuries (Bernier & Ponge 1994).

Vulnerability of ecosystem services to perturbations may be dependent on the scales at which processes and the services operate, and their comparative sizes. In theory, four different situations may occur depending on the various combination of short or large scale of either component:

1. *Short scale for processes and short scale for services.* In this case, the output of the process directly affects the service. This is the case for mineralization of nutrients by microbial digestion. In such a situation, any direct damage to the decomposer community or impairment of its activity has immediate consequences for the service. Vulnerability is high, although rapid reversibility may be expected as soon as processes are reinitiated.

2. *Short scale for processes and large scale for services.* Here, a disturbance is temporary in nature and does not impair the provision of service. This is the case for water infiltration and retention in soils. Temporary interruption of activities of those invertebrates and roots that create or rejuvenate aggregates of different sizes does not have important consequences for soil hydraulic properties (Alegre et al. 1996). Soil aggregates of biological origin, once stabilized, can be very resistant structures that last for long periods in the absence of drastic direct physical impacts. In this situation, vulnerability is minimal (Blanchart et al. 1999).

3. *Large scale for processes and short scale for services.* Disturbance of a large-scale process will impair the services for a long period. This is the case of salinization, in which soil translocation processes result in the migration of salts toward the soil surface (Luna Guido et al. 2000). In saline soils, biological processes are severely limited. These long-term processes, when achieved, may impair the service for long periods and even be irreversible in the order of decades. Vulnerability is high, and restoration of the service will depend on changes in biological communities by such processes as colonization or local adaptation and genetic selection, which can be very slow.

4 *Large scale for processes and large scale for services.* Vulnerability can now be considerable and restoration of ecosystem services is very slow. This is the case for erosion processes that may affect large portions of soils that can be restored only after a long period of soil formation through pedogenetic processes (Lal 1984). All services linked to the presence of a thick soil with well-differentiated horizons will be severely affected by such an event.

Extrinsic and Intrinsic Determinants of Vulnerability

Soil organisms vary considerably in their susceptibility to global change, and even the same taxonomic and/or functional group may vary in its response according to the nature, extent, and frequency and intensity of perturbation (Wall et al. 2001). Thus, the vulnerability of individual components of the soil fauna is context dependent.

Determinants of vulnerability are wide ranging, encompassing both intrinsic and extrinsic factors. Many of these are intuitive, and empirical assessments of vulnerability to a particular driver of global change are surprisingly rare (but see Ruess et al. 1999). One conspicuous gap in knowledge lies in the significance of species interactions in determining the vulnerability of specific organisms or the assemblages they constitute.

Among the intrinsic determinants of vulnerability is a suite of life-history traits including body size, life-cycle longevity, and host plant and habitat specificity. Soil organisms are more vulnerable to perturbations if they are large bodied (Wardle 1995; Eggleton et al. 1998). Within this domain, surface-dwelling and relatively sedentary taxa are particularly prone to changes in soil moisture and texture (e.g., Brown et al. 2001). The life-history traits of high dispersal and migratory ability tend to counteract local extinctions, though the efficacy of these traits in stemming population decline is dependent on local source pools and inherent landscape diversity (Warren et al. 2001).

Longer-lived taxa are particularly vulnerable to perturbation, since they have less chance to recover than multivoltine (producing several generations in one year) species. Within the invertebrates, soil-dwelling species are characterized by having longer generation times than their aboveground counterparts (Andersen 1987; Brown & Gange

1990), a trait that undoubtedly increases their susceptibility to change. The implications of the reduced mobility and dispersal and longer life spans of soil organisms are very apparent when trying to restore the lost diversity of a system. In the restoration of ex-arable land, it has been shown that the colonization and establishment of soil biodiversity lags behind that of aboveground organisms (Korthals et al. 2001). Organisms that are either food or habitat specialists are more susceptible to even minor changes in their resources. Strict host plant or vegetation structure specialist species may respond negatively to changes in the occurrence (Cherrill & Brown 1990) or quality of their resource (Masters et al. 1993), whereas generalist species may switch hosts or move to adjacent habitats with global change (Bale et al. 2002). Both spatial and temporal synchrony with their required resource is a pivotal requirement of specialist species, with even slight asynchrony potentially causing extinction. Specialism is therefore a key indicator of the extent and spatial scale of vulnerability to global change.

An organism's vulnerability is also related to its ability to withstand change, in terms of "flexibility" (resilience) or "rigidity" (resistance). These attributes will vary intrinsically, but also in response to local conditions. Species may be unchanged, susceptible and become locally extinct (depending on recruitment), opportunistic and increase their performance in the new environment, or elastic and change in the short term but then recover to their former level (Brown et al. 2001).

Extrinsic determinants of vulnerability reflect habitat characteristics, such as the diversity of habitat and landscape. It is a common assumption that mosaic landscapes (roughly defined as more than one land use or land cover per hectare) and the species they support are less vulnerable to perturbation and more likely to recover quickly from disturbance or depletion, but this is unproven. Small fragments of pristine ecosystems (for example patches of primary tropical forest as small as 0.25 ha) may retain high biodiversity, at least over the medium term, after isolation (de Souza & Brown 1994; Eggleton et al. 2002), but there are few measurements of associated processes and services, and long-term effects are largely unknown.

Mechanisms Underlying Vulnerability

Soil organisms and processes can be highly responsive to global change phenomena, and a variety of mechanisms are responsible. Below, we outline three main categories of global change drivers, which we consider briefly.

Belowground Responses to Different Global Change Drivers

There are those drivers that result from changes in atmospheric properties (e.g., CO_2 enrichment, N deposition, climate change), those that arise from direct manipulation of land (e.g., land use change, intensification), and those that involve shifts in organism presence or abundance (e.g., extinctions, invasions, outbreaks).

Of the human-induced changes in atmospheric composition, the enhancement of CO_2 concentration is arguably the most pervasive. Given the high levels of CO_2 in the soil, atmospheric CO_2 enrichment is unlikely to affect the soil biota directly, and effects of enrichment on the decomposer community are likely in the first instance to be plant driven. This can occur at the level of both the individual plant and the plant community. At the whole-plant level, CO_2 enrichment can promote soil organisms through increasing resource quantity, for example, by promoting NPP (Körner & Arnone 1992) and rhizodeposition (Paterson et al. 1996). Further, CO_2 enrichment can alter litter quality, frequently enhancing litter C:N ratios, although both positive and negative effects of CO_2 enrichment on decomposability of litters have been reported (Franck et al. 1997; Coûteaux et al. 1999). Elevation of CO_2 concentrations can also alter the role that soil fauna play in decomposition, since these organisms are more important in catalyzing breakdown of litter of poorer quality (Coûteaux et al. 1991). At larger spatial scales, CO_2 enrichment is also likely to alter plant community composition, frequently favoring plant species or functional types with faster growth and higher litter quality over slower growing plants (Collatz et al. 1998; Herbert et al. 1999). This is, in general, likely to favor decomposer activity.

Effects of CO_2 enrichment on the composition of the plant community directly alters the community of root pathogens and root herbivores, due to changed plant-soil feedback (Bever 1994). Trade-offs between plant growth rate and plant defense against herbivores and pathogens (van der Meijden et al. 1988) result in more specialist root herbivores and root pathogens following CO_2 enrichment. However, effects of root herbivores and pathogens are also influenced by factors other than resource availability and resource quality—such as the recognition of plant roots by pathogens, herbivores, and their natural antagonists—and by the ability of plants to culture and enumerate the antagonists of their enemies (van der Putten 2003). Litter with a high C:N ratio may not favor microbial decomposition, but it could still enhance specific microbial antagonists (Hoitink & Boehm 1999). Therefore, exposure of plant communities to CO_2 enrichment may lead to changed root pathogen and herbivore communities, reduced or more specialized activity of these root organisms, and to changed host recognition or altered exposure to the natural enemies of the herbivores and pathogens belowground.

Atmospheric N deposition also has positive effects on those soil biota that depend on plant derived resource quality through promoting NPP, and unlike CO_2 enrichment it often causes plants to produce litter with lower C:N ratios. Further, N deposition can have important belowground effects by altering the composition of the plant community, usually by favoring plant species that are adapted to fertile situations and that produce high-quality litter (Aerts & Berendse 1988). However, direct effects of N deposition on soil microbes can either promote or inhibit soil processes; chronic N deposition can either promote litter decay through enhancing microbial cellulolytic activity, or suppress it by inhibiting ligninolytic enzyme activity (Carreiro et al. 2000). Effects of N deposition on the root herbivore and pathogen community will be mediated primarily through the plant community, but changes in the physico-chemical soil properties may also exert direct effects on these soil organisms.

Global warming and associated climate change events resulting from atmospheric CO_2 enrichment can influence soil organisms directly, although the main effects of climate change on soil biota are again likely to be indirect and driven by plant responses. At the whole-plant level, increased temperature promotes the kinetics of nutrient mineralization, plant nutrient uptake, and NPP (Nadelhoffer 1992) that should in turn promote decomposer organisms. At the level of the plant community, elevated temperature has the capacity to promote plant functional types with either superior (Pastor & Post 1988; Starfield & Chapin 1996) or poorer litter quality (Hattersley 1983; Harte & Shaw 1995). In this light, the implications of climate-driven vegetation change for the decomposer subsystem are likely to be context specific. Global warming may also disrupt natural communities because of different dispersal and migration capacities of individual species (Warren et al. 2001). Migration of plant species without their natural root pathogens and herbivores may lead to enhanced abundance in the new territories, due to the escape from specific natural enemies and the presence of relatively aspecific mutualistic symbionts, for example, mycorrhizal fungi (Klironomos 2002).

Direct use of land by humans for production of food and fiber arguably has the greatest impact on terrestrial ecosystems of all global change phenomena. At the broad scale, forest, grassland, and arable systems differ tremendously in their functional composition of vegetation as well as disturbance regimes. This in turn has important implications for the composition, abundances, and activities of the soil organisms present; generally cropping systems contain lower levels of many components of soil fauna, microbial biomass, and organic matter than do comparable areas under forest or grassland (Lavelle 1994; Wardle 2002). Conversion of land to agriculture often has adverse effects on the performance of the decomposer subsystem, and artificial inputs are therefore required to substitute for services provided by soil biota. Agricultural intensification also influences the decomposer subsystem (De Ruiter et al. 1993; Giller et al. 1997). For example, agricultural tillage favored bacterial-based energy channels over fungal-based channels (Hendrix et al. 1986), promoted small-bodied soil animals relative to large-bodied ones (Wardle 1995), and altered the relative contribution of different subsets of the soil biota to litter decomposition (Beare et al. 1992). Comparable effects appear to result from intensification of forest management (Blair & Crossley 1988; Sohlenius 1996).

Another major land use change is extensification, or even the complete abandonment of production on arable land and grassland to conserve, or restore, former biodiversity (van der Putten et al. 2000). However, there has been little work on the restoration of diversity belowground. Soil communities respond more slowly to changes imposed by land abandonment than communities above ground (Korthals et al. 2001), and responses are often idiosyncratic (Hedlund et al. 2003). Therefore, besides dispersal limitations of many organisms to restoration areas (Bakker & Berendse 1999), slow development rates of the belowground community may be an important controlling factor of ecosystem services and goods provided by these newly developing restored ecosystems.

Alien species are most likely to alter community and ecosystem properties when they show large functional differences to the native species of the community being invaded. The functional attributes of most alien species do not differ greatly from those of native biota (Thompson 1995), but differences, when they do exist, can cause profound implications for both the aboveground and belowground components of the ecosystem. For example, invasion of the N-fixing shrub *Myrica faya* into Hawaiian montane forests lacking nitrogen fixing plants has important effects on ecosystem N inputs (Vitousek & Walker 1989). Introduction of deer and goats into New Zealand rainforests, which lack native browsing mammals, has caused large shifts in the soil food web composition and diversity, and in ecosystem C sequestration (Wardle et al. 2001). Invasion of European earthworms into those North American forests that lack a native earthworm fauna has been shown to alter soil microbes, fauna, and supply of plant-available nutrients from the soil (Hendrix & Bohlen 2002). Human-induced extinctions of organisms may also be of functional importance, but only in instances in which the lost species plays an important functional role. Historical examples include probable large-scale alteration of soil processes following vegetation change caused by extinctions of dominant megaherbivore species, for example in Siberia (Zimov et al. 1995) and Australia (Flannery 1994).

Soil organisms and processes are capable of showing a variety of responses to drivers of global change, and the nature of these responses is likely to be context specific. Different global change drivers do not operate independently of one another; a given ecosystem is likely to be affected by several drivers operating simultaneously. Interactions between different global change drivers (e.g., between CO_2 enrichment and N deposition [Lloyd 1999]or between invasive plants and CO_2 enrichment [Smith et al. 2000]) may have important, though largely unrealized, implications for the decomposer subsystem and for the interactions between plants, root pathogens, and symbiotic mutualists (Richardson et al. 2000). This will in turn affect those ecosystem services driven by soil organisms. Importantly, soil biota do not function in isolation, and ecosystems are driven by feedbacks between the aboveground and belowground biota (van der Putten et al. 2001; Wardle 2002; Bardgett & Wardle 2003).

Relative Vulnerability of Different Biotic Components of Communities

Subsets of the soil biota may differ considerably in terms of how they are affected by global change phenomena. Of these phenomena, the best understood for the soil biota is land use change, including intensification of land management practices. The degree

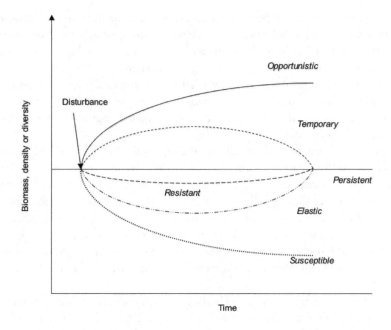

Figure 5.1. Different response dynamics of soil biota to disturbance (from Brown et al. 2001). The effect of a disturbance can result in changes to soil biomass, density, or diversity with very different results over time. This will affect the ecosystem services provided by the biota.

of impact of global change phenomena, such as land use change, is dependent on the organism's ability to withstand change, the organism's and ecosystem's resilience and resistance to the imposed changes, and the extent of the changes/disturbance imposed (difference from original environment). These can generally follow the different response strategies shown in Figure 5.1. Some organisms are susceptible to certain land management practices and become locally extinct, while others are opportunistic and take advantage of the modified conditions to increase their abundance, biomass, and activity. For example, the conversion of Amazonian rainforest to pastures north of Manaus led to the elimination of many (morpho-) species and groups of macrofauna, while one species of earthworm (*Pontoscolex corethrurus*) became the dominant soil macroorganism, reaching biomass values of up to 450 kg ha^{-1} (Barros 1999). This led to the progressive accumulation of macro-aggregates (worm castings) on the soil surface, dramatically decreasing soil macroporosity down to a level equivalent to that produced by heavy machinery, rendering it anaerobic and increasing methane emission and de-nitrification (Chauvel et al. 1999).

Some organisms may increase or decrease in numbers and biomass for only a short period (temporary or elastic) but then return to predisturbance proportions, while others remain unchanged or only slightly unchanged (persistent or resistant). For instance, if new land-use practices imposed maintain enough similarities to the previous ecosystem (e.g., conversion of native grass savanna to pastures), many soil organisms may resist the change, while some of those negatively affected in the land-preparation and conversion phase (tillage and seeding or transplanting) may eventually recover in the new system once proper soil cover and plant organic matter inputs are re-established (Jiménez & Thomas 2001).

Conversion from forest to agriculture usually results in an overall reduction in microarthropod populations (Crossley et al. 1992), but the response dynamics of component taxa varies. Continuous cultivation, rotations, monoculture, and application of pesticides soon eliminate species susceptible to damage, desiccation, and destruction of their microhabitats, especially those with a life cycle longer than one year, such as many oribatid mites. In contrast, practices such as drainage, irrigation, manuring, and fertilizer use encourage seasonal multiplication of species of prostigmatid mites and Collembola, and their predators, mesostigmatid mites (Crossley et al. 1992).

Several studies reveal that increases in the intensity of agriculture lead to reductions in the diversity of soil biota (Siepel 1996; Yeates et al. 1997). This diversity loss is not a random process. For instance, Siepel (1996) showed that declines in the diversity of soil microarthropods with increasing agricultural intervention were accompanied by dramatic shifts in the life-history characteristics and feeding guilds of the community. Loss of diversity in low-input agricultural systems was explained by the disappearance of drought-intolerant species because low-input grasslands are cut in summer, thereby increasing the chance of drought in the litter layer. However, the loss of species in high-input grassland was explained by the elimination of fungal-feeding grazers that were replaced by opportunistic bacterial-feeders. Moreover, abandoned high-input sites still lacked fungal-feeding mites, even after 20 years of management for nature conservation, due to the low population growth and dispersal rate of these species (Siepel 1996).

Effects of land-use practices on soil biota in turn exert profound effects on key ecosystem processes that they perform, and ultimately on the delivery of ecosystem services. Enhanced disturbance regimes tend to favor the bacterial-based energy channel of the soil food web over the fungal-based channel, and management practices which are known to favor the bacterial channel include conventional (vs. non) tillage (Hendrix et al. 1986), nitrogen fertilization (Ettema et al. 1999) and forest clear-cutting (Sohlenius 1996). Domination of soil food webs by bacterial energy channels leads to greater short-term mineralization rates of carbon and nutrients, leading to greater net losses of soil organic matter and reduced retention of nutrients in the soil in the longer term. The tendency of agricultural intensification practices to favor soil animals of smaller body sizes probably also has similar effects on ecosystem nutrient losses (Wardle 1995).

Therefore, effects of land-use practices on the composition of soil biota may have important flow-through effects on key ecosystem services provided by soil organisms, in particular those relating to the maintenance of soil fertility and plant-available nutrient supply.

Unfortunately, given the overwhelming diversity of soil organisms and their functions and interactions in soils, information on the response dynamics of various groups/taxa and species of soil organisms to land-use change and its possible effects on soil function is not available for many sites/land uses. Furthermore, there can also be important differences in the effects on the same organism/function to forces of change, depending on local climate or soil conditions. Processes occurring locally or influenced by organisms at a small scale are less likely to be influenced by change at a large scale, unless it undermines their populations or ability to maintain activity. However, cascading effects of loss of a particular organism or function within the ecosystem could have consequences far beyond the small scales at which they were initially operating.

Assessment of Vulnerability

Arable, grassland, and forest ecosystems are managed primarily for the purpose of providing material goods such as forage, food, and fiber. However, terrestrial ecosystems also provide a range of other goods and services, notably through the improvement of environmental quality (e.g., water purification, flood and erosion control, atmospheric regulation), recreational and amenity values, provision of habitats for species and conservation of biodiversity, and mitigation of anthropogenic CO_2 enrichment through sequestering carbon. The soil biota play an important role in the delivery of all of these services. This is in part because the above- and belowground components of terrestrial ecosystems are inextricably linked. Therefore, any global change agent that affects the soil community will affect not only the ability of the soil to provide services that are directly driven by the soil biota (e.g., soil carbon sequestration, prevention of leaching of nutrients), but also those that are driven by the plant community.

To illustrate the nature and mechanistic basis of vulnerability of ecosystem goods and services driven by soil biota to global change, we now present some examples. In Chapter 2, an assessment was made of the importance of contributions of soil biotic and abiotic factors to the delivery of ecosystem services in representative temperate arable tilled, grassland, and forest ecosystems. These were quantified on a scale of * (unimportant) to *** (highly important). In the present assessment, we considered, for each of these ecosystems, the vulnerability of each of those ecosystem services identified in Chapter 2 to three scenarios of global change: global change agents that (1) operate over large scales via the atmosphere (drought, resulting from climate change), (2) operate at the landscape scale (change of the ecosystem to a new land use), and (3) operate directly through individual species effects (invasions by alien species with radically different functional attributes to that of the resident species). For each of these scenarios, we considered how each ecosystem service is modified by the global change agents, how abiotic and biotic drivers of that service are modified, and how the role of soil biota in providing that service is altered. We used our understanding of current knowledge in the published literature, and assigned semi-quantitative scores. Specifically, we assessed the extent to which the ability of the soil biota to provide each good or service was vulnerable to each global change agent for each ecosystem type, assigning a score to each ability by using a scale of 0 to 3 (with increasing values indicating greater vulnerability). We also assessed the net impact of the global change factor (positive, neutral, or negative) on the delivery of the good or service.

The Quantification Exercise

Arable tilled ecosystems represent the most biologically simple of the three ecosystem types that we evaluated. The three perturbations that we considered were (1) invasive species: root parasites; (2) climate change: drought; and (3) land-use change: to forest. The quantification exercise (Table 5.A1 in the appendix on page 121) revealed high vulnerability of the primary ecosystem services provided by arable systems (food and fiber production) to both the invasive species and climate change scenarios, and also high vulnerability of fiber production to land-use change. However, there were also several instances in which secondary services (e.g., water quality, flood and erosion control, habitat provision, recreational values, carbon sequestration) were vulnerable to specific global change phenomena. For the majority of cases, negative impacts of both the invasive root pathogen and drought on the ecosystem good or service under consideration was predicted. This is because most of these goods and services are maximized by plant productivity, and both scenarios operate to reduce productivity, as well as reduce the ability of the soil biota to maximize productivity. In contrast, the land-use change scenario—that is, afforestation—had positive effects on many ecosystem services (Table 5.A1 in the appendix on page 121). This relates to forests representing a more complex, integrated ecosystem than arable tilled systems, and one in which the soil biota, rather than artificial inputs, plays a much greater role in the delivery of ecosystem services. Afforestation is therefore likely to enhance the role that soil biota plays in the delivery of such varied services as ecosystem carbon sequestration, nutrient retention, atmospheric regulation, habitat provision, and flood and erosion control.

Temperate grasslands are usually dominated by perennial plant species, and therefore typically represent a more complex, lower input ecosystem type than do arable ecosystems. The three perturbations that we considered were (1) invasive species: a generic invasive plant (weed) species; (2) climate change: drought; and (3) land-use change: to arable land. The quantification exercise (Table 5.A2 in the appendix on page 126) again predicted strong effects of global change phenomena on material goods provided by the ecosystem, such as food and fiber, and these effects were frequently negative. Global change effects on these goods arise through their influences on plant community composition and productivity, the ability of soil food web organisms to maximize this productivity, and linkages between the plant and soil community. This can involve important shifts from dominance by fungal based food webs and soil animals with large body sizes to bacterial based food webs and small bodied soil animals, especially in the case of the land-use change scenario. Global change phenomena also have wide ranging (and in the majority of cases, negative) effects on environmental services provided by the soil biota, including maintenance of water quality, habitat provision, bioremediation, recreation, carbon sequestration, and atmospheric regulation of gases (Table 5.A2 in the appendix on page 126). The effects of invasive plant and climate change on these types of services may arise through their adverse effects on plant productivity and soil biotic activity. Meanwhile, land-use change may affect these services through the grassland changing to a more biologically simplistic system in which the importance of the soil biota relative to that of artificial inputs in providing services diminishes.

Forest ecosystems represent the most complex and biologically organized of the three ecosystem types that we considered. The three perturbations evaluated in this case were (1) invasive species: exotic earthworms; (2) climate change: drought; and (3) land-use change: deforestation. The quantification exercise (Table 5.A3 in the appendix on page 132) revealed likely effects of all three agents on material goods provided by the forest, notably timber and wood-based products. Although deforestation directly and obviously impairs the ability of the forest to produce wood, there are also mechanisms through which drought and earthworm invasion may influence performance of both the aboveground and belowground subsystems that ultimately affect timber production (Table 5.A3 in the appendix on page 132). Forests, through virtue of being less disturbed by humans than grassland or arable systems, frequently have far greater recreation and biodiversity conservation values; the ability of the plant-soil system to provide these values are potentially indirectly responsive to both deforestation and drought (Table 5.A3 in the appendix on page 132). Environmental services provided by forests to which the soil biota contributes (e.g., water retention, erosion control, carbon sequestration, and regulation of atmospheric gases) are all maximized by forest stands with high biomass, which are in turn maintained by both decomposer food web activity and mycorrhizal associations. Ultimately, the type of global change agent operating will determine the direction of responses of ecosystem services; of the examples presented here, invasion of earthworms may increase plant productivity and the role of soil biota in providing these services, while drought and forest clearance may be expected to have generally detrimental effects.

Conclusions

There are many determinants of vulnerability of soil biota and the services that they provide to global change. The overarching determinant is spatial and temporal scale; global change phenomena are simultaneously manifested at a range of scales, and can affect soil biota at each of these scales. Soil organisms and the processes that they regulate also function at several scales, and this in turn results in the effects of global change on services provided by the soil biota being inherently scale dependent. Further, a range of extrinsic and intrinsic factors influences the vulnerability of services delivered by the soil biota to global change. Among these are life-history traits that determine the resilience and resistance of organism populations to perturbations, including those created through global change, and therefore the processes driven by these organisms. There are numerous mechanisms through which global change phenomena can affect soil biota, and these have varied and complex effects on both the organisms themselves and the services that they regulate. The direction of these effects depends upon the global change phenomenon considered, spatial and temporal scale, and the community composition of both the aboveground and belowground biota (Wardle et al. 2004). A recurrent theme is the overarching role of linkages between the aboveground and belowground subsystems: these subsystems do not operate in isolation but are instead mutually dependent upon one another. Global change factors that directly affect organisms on one side of the aboveground-belowground interface will therefore promote feedbacks through their indirect effects on organisms on the other side of the interface.

The quantification exercise (Tables 5.A1–5.A3) serves to reinforce the important role that soil organisms have in driving ecosystem services, as well as the extent to which they are affected by global change. Although specific entries and scores can be debated, and it is recognized that this exercise has the usual limitations of any survey based on expert opinion, the tables nevertheless provide clear evidence that the soil biota play an important role in the delivery of a range of ecosystem services in very different ecosystem types, and that this role can be affected (either positively or negatively) by a spectrum of global change phenomena. Due to the many linkages of aboveground and belowground subsystems, soil biota probably play at least some role (however indirectly) in every terrestrial ecosystem service that has a biological component. Ultimately, if we are to understand better how ecosystems deliver goods and services upon which we depend and how these can be managed under scenarios of global change, then it is imperative that we recognize the considerable contribution that soil biota make to these services and their response to global change phenomena.

Literature Cited

Aerts, R., and F. Berendse. 1988. The effect of increased nutrient availability on vegetation dynamics in wet heathland. *Vegetatio* 76:63–69.

Alegre, J.C., B. Pashanasi, and P. Lavelle. 1996. Dynamics of soil physical properties in a low input agricultural system inoculated with the earthworm *Pontoscolex corethrurus* in the Amazon region of Peru. *Soil Science Society of America Journal* 60:1522–1529.

Andersen, D.C. 1987. Below-ground herbivory in natural communities: A review emphasizing fossorial animals. *Quarterly Review of Biology* 62:261–286.

Bakker, J.P., and F. Berendse. 1999. Constraints in the restoration of ecological diversity in grassland and heathland communities. *Trends in Ecology and Evolution* 14:63–68.

Bale, J.S., G.J. Masters, I.D. Hodkinson, C. Awmack, T.M. Bezemer, V.K. Brown, J. Butterfield, A. Buse, J.C. Coulsen, J. Farrar, J.E.G. Good, R. Harrington, S. Hartley, T.H. Jones, R.L. Lindroth, M.C. Press, I. Symrndioudis, A.D. Watt, and J.B. Whittaker. 2002. Herbivory in global climate change research: Direct effects of rising temperature on insect herbivores. *Global Change Biology* 8:1–16.

Bardgett, R.D., J.M. Anderson, V. Behan-Pelletier, L. Brussaard, D.C. Coleman, C. Ettema, A. Moldenke, J.P. Schimel, and D.H. Wall. 2002. The role of soil biodiversity on hydrological pathways and the transfer of materials between terrestrial and aquatic ecosystems. *Ecosystems* 4:421–429.

Bardgett, R.D., and D.A. Wardle. 2003. Herbivore mediated linkages between aboveground and belowground communities. *Ecology* 84:2258–2268.

Barros, M.E. 1999. Effet de la macrofaune sur la structure et les processus phisiques du sol de pâturages dégradés d'amazonie. PhD Thesis, Université Paris VI.

Beare, M.H., R.W. Parmelee, P.F. Hendrix, W.X. Cheng, D.C. Coleman, and D.A. Crossley. 1992. Microbial and faunal interactions and effects on litter nitrogen and decomposition in agroecosystems. *Ecological Monographs* 62:569–591.

Bernier, N., and J.F. Ponge. 1994. Humus form dynamics during the sylvogenic cycle in a mountain spruce forest. *Soil Biology and Biochemistry* 26:183–220.

Bever, J.D. 1994. Feedback between plants and their soil communities in an old field community. *Ecology* 75:1965–1977.

Bingham, I.J. 2001. Soil-root-canopy interactions. *Annals of Applied Biology* 138:243–251.

Blair, J.M., and D.A. Crossley. 1988. Litter decomposition, nitrogen dynamics and litter microarthropods in a southern Appalachian hardwood forest 8 years following clearcutting. *Journal of Applied Ecology* 25:683–698.

Blanchart, E., A. Albrecht, J. Alegre, A. Duboisset, B. Pashanasi, P. Lavelle, and L. Brussaard. 1999. Effects of earthworms on soil structure and physical properties. In: *Earthworm Management in Tropical Agroecosystems*, edited by P. Lavelle, L. Brussaard, and P. Hendrix, pp. 139–162. Wallingford, UK, CAB International.

Branson, F.A., G.F. Gifford, K.G. Renard, and R.F. Hadley. 1981. *Rangeland Hydrology*. Dubuque, Iowa, Kendall/Hunt Publishing Company.

Brown, G.G., D.E. Bennack, A. Montanez, A. Braun, and S. Bunning. 2001. Soil biodiversity portal. In: *Conservation and Management of Soil Biodiversity and Its Role in Sustainable Agriculture*. Rome, Italy, Food and Agriculture Organization of the United Nations, http://www.fao.org/ag/AGL/agll/soilbiod/default.htm.

Brown, V.K., and A.C. Gange. 1990. Insect herbivory below ground. *Advances in Ecological Research* 20:1–58.

Buljovic, Z., and C. Engels. 2001. Nitrate uptake ability by maize roots during and after drought stress. *Plant and Soil* 229:125–135.

Burke, I.C, C.M. Yonker, W.J. Parton, C.V. Cole, K. Flach, and D.S. Schimel. 1989. Texture, climate, and cultivation effects on soil organic matter content in U.S. grassland soils. *Soil Science Society of America Journal* 53:800–805.

Carreiro, M.M., R.L. Sinsabaugh, D.A. Repert, and D.F. Parkhurst. 2000. Microbial enzyme shifts explain litter decay responses to simulated nitrogen deposition. *Ecology* 81:2359–2365.

Chauvel, A., M. Grimaldi, E. Barros, E. Blanchart, M. Sarrazin, and P. Lavelle. 1999. Pasture degradation by an Amazonian earthworm. *Nature* 389:32–33.

Cherrill, A.J., and V.K. Brown. 1990. The habitat requirements of adults of the Wartbiter *Decticus verrucivorus* (L.) (Orthoptera: Tettigoniidae) in southern England. *Biological Conservation* 53:145–157.

Collatz, G.J., J.A. Berry, and J.S. Clark. 1998. Effects of climate and atmospheric CO_2 partial pressure on the global distribution of C_4 grasses: Present, past and future. *Oecologia* 114:441–454.

Coûteaux, M.M., C. Kurz, P. Bottner, and A. Raschi. 1999. Influence of increased atmospheric CO_2 concentration on quality of plant material and litter decomposition. *Tree Physiology* 19:301–311.

Coûteaux, M.M., M. Mousseau, M.L. Celerier, and P. Bottner. 1991. Increased atmospheric CO_2 and litter quality: Decomposition of sweet chestnut leaf litter with animal foodwebs of different complexities. *Oikos* 61:53–64.

Crossley, D.A., Jr., B.R. Mueller, and J.C. Perdue. 1992. Biodiversity of microarthropods in agricultural soils: Relations to processes. *Agriculture, Ecosystems and Environment* 40:37–46.

Decaëns, T. 2000. Degradation dynamics of surface earthworm casts in grasslands of the eastern plains of Colombia. *Biology and Fertility of Soils* 32:149–156.

Del Grosso, S.J., W.J. Parton, A.R. Mosier, D.S. Ojima, C.S. Potter, W. Borken, R. Brumme, K. Butterbach-Bahl, P.M. Crill, K. Dobbie, and K.A. Smith. 2000. General CH_4 oxidation model and comparisons of CH_4 oxidation in natural and managed systems. *Global Biogeochemical Cycles* 14:999–1019.

De Ruiter, P.C., J.C. Moore, K.B. Zwart, L.A. Bouwman, J. Hassink, J. Bloem, J.A. Devos, J.C.Y. Marinissen, W.A.M. Didden, G. Lebbink, and L. Brussaard. 1993. Simulation of nitrogen mineralization in the below-ground food webs of two winter wheat fields. *Journal of Applied Ecology* 30:95–106.

de Souza, O.F.F., and V.K. Brown. 1994. Effects of habitat fragmentation on Amazonian termite communities. *Journal of Tropical Ecology* 10:197–206.

Eggleton, P., D.E Bignell, S. Hauser, L. Dibod, L. Norgrove, and B. Madong. 2002. Termite diversity across an anthropogenic disturbance gradient in the humid forest zone of West Africa. *Agriculture, Ecosystems and Environment* 90:189–202.

Eggleton, P., R.G. Davies, and D.E. Bignell. 1998. Body size and energy use in termites (Isoptera): The responses of soil feeders and wood feeders differ in a tropical forest assemblage. *Oikos* 81:525–530.

Elliott, P.W., D. Knight, and J.M. Anderson. 1991. Variables controlling denitrification from earthworm casts and soil in permanent pastures. *Biology and Fertility of Soils* 11:24–29.

Ettema, C., R. Lowrance, and D.C. Coleman. 1999. Riparian soil response to surface nitrogen input: The indicator potential of free-living soil nematode populations. *Soil Biology and Biochemistry* 31:1625–1638.

Ferrier, R.C., A.C. Edwards, J. Dutch, R. Wolstenholme, and D.S. Mitchell. 1996. Sewage sludge as a fertilizer of pole stage forests: Short-term hydrochemical fluxes and foliar response. *Soil Use and Management* 12:1–7.

Flannery, T.F. 1994. *The Future Eaters*. Melbourne, Australia, Reed Books.

Franck, V.M., B.A. Hungate, F.S. Chapin III, and C.B. Field. 1997. Decomposition of litter produced under elevated CO_2: Dependence on plant species and nutrient supply. *Biogeochemistry* 36:223–237.

Garbaye, J. 2000. The role of ectomycorrhizal symbiosis in the resistance of forests to water stress. *Outlook on Agriculture* 29:63–69.

Giller, K.E., M.H. Beare, P. Lavelle, A.M.N. Izac, and M.J. Swift. 1997. Agricultural intensification, soil biodiversity and ecosystem function. *Applied Soil Ecology* 6:3–16.

Haimi, J., and M. Einbork. 1992. Effects of endogenic earthworms on soil processes and plant-growth in coniferous forest soil. *Biology and Fertility of Soils* 13:6–10.

Harte, J., and R. Shaw. 1995. Shifting dominance within a montane vegetation community: Results of a climate-warming experiment. *Science* 267:876–880.

Hattersley, P.W. 1983. The distribution of C_3 and C_4 grasses in Australia in relation to climate. *Oecologia* 57:113–128.

Hedlund, K., I. Santa Regina, W.H. van der Putten, G.W. Korthals, T. Díaz, J. Leps, S. Lavorel, J. Roy, D. Gormsen, S.R. Mortimer, P. Smilauer, M. Smilauerová, C. Van Dijk, V.K. Brown, and C. Rodríguez Barrueco. 2003. Plant species diversity, plant biomass and responses of the soil community on abandoned land across Europe: Idiosyncrasy or above-belowground time lags. *Oikos* 103:45–58.

Hendrix, P.F., and P.J. Bohlen. 2002. Exotic earthworm invasions in North America: Ecological and policy implications. *BioScience* 52:801–811.

Hendrix, P.F., R.W. Parmelee, D.A. Crossley, D.C. Coleman, E.P. Odum, and P.M. Groffman. 1986. Detritus food webs in conventional and no-tillage agroecosystems. *BioScience* 36:374–380.

Herbert, D.A., E.B. Rastetter, G.R. Shaver, and G.I. Ågren. 1999. Effects of plant growth characteristics on biogeochemistry and community composition in a changing climate. *Ecosystems* 2:367–382.

Hirsch, S.A., and J.A. Leisch. 1998. Impact of leafy spurge on post-conservation reserve program land. *Journal of Range Management* 51:614–620.

Hoitink, H.A.J., and M.J. Boehm. 1999. Biocontrol within the context of soil microbial communities: A substrate-dependent phenomenon. *Annual Review of Phytopathology* 37:427–446.

Hunt, H.W., E.T. Elliott, and D.E. Walter. 1989. Inferring trophic transfers from pulse-dynamics in detrital food webs. *Plant and Soil* 115:247–259.

Ineson, P., J. Dutch, and K.S. Killham. 1991. Denitrification in a Sitka spruce plantation and the effect of clear-felling. *Forest Ecology and Management* 44:77–92.

Jacobs, J.S., and R.L. Sheley. 1998. Observation: Life history of spotted knapweed. *Journal of Range Management* 51:665–673.

Jiménez, J.J., and R.J. Thomas. 2001. *Nature's Plow: Soil Macroinvertebrate Communities in the Neotropical Savannas of Colombia*. Cali, Colombia, Centro Internacional de Agricultura Tropical (CIAT).

Klironomos, J.N. 2002. Feedback with soil biota contributes to plant rarity and invasiveness in communities. *Nature* 417:67–70.

Körner, C., and J.A. Arnone III. 1992. Response to elevated carbon dioxide in artificial tropical ecosystems. *Science* 257:1672–1675.

Korthals, G.W., P. Smilauer, C. Van Dijk, and W.H. van der Putten. 2001. Linking above and below-ground biodiversity: Abundance and trophic complexity in soil as a response to experimental plant communities on abandoned arable land. *Functional Ecology* 15:506–514.

Kourtev, P.S., J.G. Ehrenfeld, and M. Haggblom. 2002. Exotic plant species alter the microbial community structure and function in the soil. *Ecology* 83:3152–3166.

Kourtev, P.S., W.Z. Huang, and J.G. Ehrenfeld. 1999. Differences in earthworm densities and nitrogen dynamics in soils under exotic and native plant species. *Biological Invasions* 1:237–245.

Lal, R. 1984. Soil erosion from tropical arable land and its control. *Advances in Agronomy* 37:187–248.

Lavelle, P. 1994. Faunal activities and soil processes: Adaptive strategies that determine ecosystem function. In: *XV ISSS Congress Proceedings, Acapulco, Mexico, Volume 1: Introductory Conferences,* pp. 189–220.

Lavelle, P. 1997. Faunal activities and soil processes: Adaptive strategies that determine ecosystem function. *Advances in Ecological Research* 27:93–132.

Lavelle, P., D. Bignell, M. Lepage, V. Wolters, P. Roger, P. Ineson, O.W. Heal, and S. Dhillion. 1997. Soil function in a changing world: The role of invertebrates as ecosystem engineers. *European Journal of Soil Biology* 33:159–193.

Lawrence, B., M.C. Fisk, T.J. Fahey, and E.R. Suarez. 2003. Influence of non-native earthworms on mycorrhizal colonization of sugar maple (*Acer saccharum*). *New Phytologist* 157:145–153.

Lloyd, J. 1999. The CO_2 dependence of photosynthesis, plant growth responses to elevated CO_2 concentrations and their interaction with soil nutrient status, II. Temperate and boreal forest productivity and the combined effects of increasing CO_2 concentrations and increased nitrogen deposition at a global scale. *Functional Ecology* 13:439–459.

Luna Guido, M.L., R.I. Beltran Hernandez, N.A. Solis Ceballos, N. Hernandez Chavez, F. Mercado Garcia, J.A. Catt, V. Olalde Portugal, and L. Dendooven. 2000. Chemical and biological characteristics of alkaline saline soils from the former Lake Texcoco as affected by artificial drainage. *Biology and Fertility of Soils* 32:102–108.

Masters, G.J., V.K. Brown, and A.C. Gange. 1993. Plant mediated interactions between above- and below-ground insect herbivores. *Oikos* 66:148–151.

Myers, N. 2002. Environmental refugees: A growing phenomenon of the 21st century. *Philosophical Transactions of the Royal Society of London: Biological Sciences.* 357:609–613.

Nadelhoffer, K. 1992. Microbial processes and plant nutrient availability in arctic soil. In: *Arctic Ecosystems in a Changing Climate. An Ecophysiological Perspective,* edited by F.S. Chapin III, R.L. Jefferies, J.F. Reynolds, G.R. Shaver, and J. Svoboda, pp. 281–300. San Diego, California, Academic Press.

Nye, P.H. 1955. Some soil-forming processes in the humid tropics. IV. The action of the soil fauna. *Journal of Soil Science* 6:73–83.

Pastor, J., and W.M. Post. 1988. Response of northern forests to CO_2-induced climate change. *Nature* 334:55–58.

Paterson, E., E.A.S. Rattray, and K. Killham. 1996. Effect of elevated atmospheric CO_2 concentration on C-partitioning and rhizosphere C-flow for three plant species. *Soil Biology and Biochemistry* 28:195–201.

Resilience Alliance. 2002. Vulnerability. http://resalliance.org/ev_en.php?ID=1119_21& IDZ=DO_TOPIC

Richardson, D.M., N. Allsopp, C.M. D'Antonio, S.J. Milton, and M. Rejmanek. 2000. Plant invasions: The role of mutualisms. *Biological Reviews* 75:65–93.

Robinson, C.H., P. Ineson, T.G. Piearce, and A.P. Rowland. 1992. Nitrogen mobilization by earthworms in limed peat soils under *Picea sitchensis*. *Journal of Applied Ecology* 29:226–237.

Ruess L., A. Michelsen, I.K. Schmidt, and S. Jonasson. 1999. Simulated climate change affecting microorganisms, nematode density and biodiversity in subarctic soils. *Plant and Soil* 212:63–73.

Rusek, J. 1985. Soil microstructures: Contributions on specific soil organisms. *Quaestiones Entomologicae* 21:497–514.

Saxena, D., and G. Stotzky. 2000. Insecticidal toxin from *Bacillus thuringiensis* is released from roots of transgenic *Bt* corn in vitro and in situ. *FEMS Microbiology Ecology* 33:35–39.

Siepel, H. 1996. Biodiversity of soil microarthropods: The filtering of species. *Biodiversity and Conservation* 5:251–260.

Smith, S.D., T.E. Huxman, S.F. Zitzer, T.N. Charlet, D.C. Housman, and J.S. Coleman. 2000. Elevated CO_2 increases productivity and invasive species success in an arid ecosystem. *Nature* 408:79–82.

Sohlenius, B. 1996. Structure and composition of the nematode fauna in pine forests under the influence of clear-cutting: Effects of slash removal and field layer vegetation. *European Journal of Soil Biology* 32:1–14.

Starfield, A.M., and F.S. Chapin III. 1996. Model of transient changes in Arctic and boreal vegetation in response to climate and land use change. *Ecological Applications* 6:842–864.

Thies, W.G. 2001. Root diseases in eastern Oregon and Washington. *Northwest Science* 75:38–45.

Thompson, K. 1995. Native and alien species: More of the same? *Ecography* 18:390–402.

Tiunov, A.V., and T.G. Dobrovolskaya. 2002. Fungal and bacterial communities in *Lumbricus terrestris* burrow walls: A laboratory experiment. *Pedobiologia* 46:595–605.

van der Meijden, E., M. Wijn, and H.J. Verkaar. 1988. Defense and regrowth: Alternative plant strategies in the struggle against herbivores. *Oikos* 51:355–363.

van der Putten, W.H. 2003. Plants defense below ground and spatio-temporal processes in natural vegetation. *Ecology* 84:2269–2280.

van der Putten, W.H., S.R. Mortimer, K. Hedlund, C. Van Dijk, V.K. Brown, J. Lep, C. Rodríguez-Barrueco, J. Roy, T.A. Díaz, D. Gormsen, G.W. Korthals, S. Lavorel, I. Santa-Regina, and P. Milauer. 2000. Biodiversity experiments at abandoned arable land along climate transects. *Oecologia* 124:91–99.

van der Putten, W.H., L.E.M. Vet, J.A. Harvey, and F.L. Wackers. 2001. Linking above- and belowground multitrophic interactions of plants, herbivores, pathogens and their antagonists. *Trends in Ecology and Evolution* 16:547–554.

Vitousek, P.M., and L.R. Walker. 1989. Biological invasion by *Myrica faya* in Hawaii: Plant demography, nitrogen fixation, ecosystem effects. *Ecological Monographs* 59:247–265.

Wall, D.H., P.V.R. Snelgrove, and A.P. Covich. 2001. Conservation priorities for soil and sediment invertebrates. In: *Conservation Biology*, edited by M.E. Soulé and G.H. Orians, pp. 99–123. Washington, DC, Island Press.

Wardle, D.A. 1995. Impact of disturbance on detritus food-webs in agro-ecosystems of contrasting tillage and weed management practices. *Advances in Ecological Research* 26:105–185.

Wardle, D.A. 2002. *Communities and Ecosystems. Linking the Aboveground and Belowground Components.* Princeton, New Jersey, Princeton University Press.

Wardle, D.A., R.D. Bardgett, J.N. Klironomos, H. Setälä, W.H. van der Putten, and D.H. Wall. 2004. Ecological linkages between aboveground and belowground biota. *Science* 304:1629–1633.

Wardle, D.A., G.M. Barker, G.W. Yeates, K.I. Bonner, and A. Ghani. 2001. Introduced browsing mammals in natural New Zealand forests: Aboveground and belowground consequences. *Ecological Monographs* 71:587–614.

Wardle, D.A., K.I. Bonner, G.M. Barker, G.W. Yeates, K.S. Nicholson, R.D. Bardgett, R.N. Watson, and A. Ghani. 1999. Plant removals in perennial grassland: Vegetation dynamics, decomposers, soil biodiversity, and ecosystem properties. *Ecological Monographs* 69:535–568.

Warren, M.S., J.K. Hill, J.A. Thomas, J. Asher, R. Fox, B. Huntley, D.B. Roy, M.G. Telfer, S. Jeffcoate, P. Harding, G. Jeffcoate, S.G. Willis, J.N. Greatorex-Davies, D. Moss, and C.D. Thomas. 2001. Rapid responses of British butterflies to opposing forces of climate and habitat change. *Nature* 414:65–69.

Westover, K.M., A.C. Kennedy, and S.E. Kelly. 1997. Patterns of rhizosphere microbial community structure associated with co-occurring plant species. *Journal of Ecology* 85:863–873.

Wiklund, K., L.O. Nilsson, and S. Jacobsson. 1995. Effect of irrigation, fertilization, and artificial drought on basidioma production in a Norway spruce stand. *Canadian Journal of Botany* 73:200–208.

Wolters, V., W.L. Silver, D.E. Bignall, D.C. Coleman, P. Lavelle, W.H. van der Putten, P. de Ruiter, J. Rusek, D.H. Wall, D.A. Wardle, J.M. Dangerfield, V.K. Brown, K.E. Giller, D.U. Hooper, O. Sala, J. Tiedje, and J.A. van Veen. 2000. Global change effects on above- and belowground biodiversity in terrestrial ecosystems: Interactions and implications for ecosystem functioning. *BioScience* 50:1089–1098.

Yeates, G.W., R.D. Bardgett, R. Cook, P.J. Hobbs, P.J. Bowling, and J. Potter. 1997. Faunal and microbial diversity in three Welsh grassland soils under conventional and organic management regimes. *Journal of Applied Ecology* 34:453–470.

Zimov S.A., V.I. Chuprynin, A.P. Oreshko, F.S. Chapin, III, J.F. Reynolds, and M.C. Chapin. 1995. Steppe-tundra transition: A herbivore-driven biome shift at the end of the Pleistocene. *American Naturalist* 146:765–794.

Appendix Table 5.A1. Vulnerability of ecosystem goods and services in arable tilled ecosystems provided by the soil biota to three agents of global change: invasive species, climate change, and land-use change.

The "service rank" (range −3 to +3) indicates the importance of the ecosystem under consideration (arable, tilled) in providing each ecosystem good and service; positive and negative values indicate positive and negative effects, respectively, of the ecosystem in providing that good or service. The importance of "biotic" and "abiotic" factors in providing each good or service ranges from unimportant (designated by *) to highly important (***). "Vulnerability" scores (range 0 to 3) relate to the vulnerability of each good or service provided by the ecosystem to each perturbation; greater values indicate greater vulnerability; text in brackets explains vulnerability score. "Net impact" refers to the direction of change in the ability of the ecosystem to provide each good or service after perturbation, and is scored as − (reduction), 0 (no change), or + (increase).

| | | | | Perturbations | | | | | |
| | Unmanaged (Arable Tilled)[1] | | | Invasive Species: Root Parasites[2] | | Climate Change: Drought[3] | | Land-Use Change: To Forest | |
Goods and Services	Service Rank	Biotic	Abiotic	Vulnerability	Net Impact	Vulnerability	Net Impact	Vulnerability	Net Impact
Food production	3	** plant breeding, roots and residues	*** climate, tillage, soil type, fertilizer	3 [can lead to destruction of crop]	−	3 [abiotic controls, especially soil properties and fallow practices, most important; reduced NPP; drought-resistant plant breeds important[5]]	−4	0 [litter system developed; annual roots replaced by perennial; increased role for soil food web; change from bacterial to fungal-based food web[6]]	0
Water quality	−2	* nitrification; pesticide degradation	*** topography; leaching; NH_4 volatilization	2 [increased application of pesticides negatively impacts water quality.]	−	0 [service is effectively cancelled.]	−	0 [improvement in quality through restoration of ecosystem engineering communities and creation of biogenic structures]	+

(continued)

Appendix Table 5.A1. (continued)

Goods and Services	Unmanaged (Arable Tilled)[1]			Perturbations					
				Invasive Species: Root Parasites[2]		Climate Change: Drought[3]		Land-Use Change: To Forest	
	Service Rank	Biotic	Abiotic	Vulnerability	Net Impact	Vulnerability	Net Impact	Vulnerability	Net Impact
Flood and erosion control[8]	−3	*** plant cover, crop type (roots)	*** topography, soil properties	0	0	3 [abiotic controls, especially fallow practices, most importan[9]; drought-resistant plant breeds important[10]]	+[7]	0 [restoration of strong biotic control through change in root architecture, mycorrhizal associations and creation of biogenic structures]	+
Fiber	1	** plant breeding, roots and residues	*** climate, tillage, soil type, fertilizer	3 [can lead to destruction of crop]	−	3 [abiotic controls, especially soil properties and fallow practices, most important; reduced NPP]	−	3 [litter system developed; annual roots replaced by perennial; increased role for soil food web; change from bacterial to fungal-based food web[6]]	+
Waste disposal/bio-remediation	3	** decomposition, co-metabolism, build up of intermediate toxic products	** volatilization, water regime, soil properties, sequestration, incorporation by machinery	1 [dependent on rates of decomposition]	− / +	3 [dependent on soil type and pre-drought conditions]	−	0 [no fundamental change in service, but change in rate and performed by different food web communities[6]]	+
Biological control[11]	−1	** rotation, GMOs[12], micro-food webs	** pesticides, tillage	1	−	1 [can cause shift from specific to generalist pests[13]]	0	0 [restoration of strong biotic control]	+

Recreation and other uses	–1	** odor, monoculture/low landscape heterogeneity	*** erosion, topography seasonally bare land; noise, air pollution, pesticides	1 [increased pesticide use]	–	2 [potential loss of landscape features through erosion, increase in bare land and salinization]		0 [dependent on forest type and management[14]]	+
Carbon sequestration	–1	* decomposition	** topography texture; nature of clay minerals; climate; soil structure	0–1 [dependent on changes in rate of decomposition]	–	0–1 [dependent on abiotic factors, irrigation, vulnerability to fire]	–/+[15]	0 [dependent on forest type[16], litter quality and quantity, humification and sequestration in biogenic structures]	+
Trace gases and atmospheric regulation	–3	** ammonification, nitrification	** moisture regime, soil structure	0 [dependent on whether other biotic factors affected]	–/+	0 [biotic activity slowed, thus decrease in production of trace gases]	–/+	0 [dependent on litter quality and quantity and nitrifiers, denitrifiers, methane oxidizers]	+
Fuel	0	NA	NA	NA	0	NA	NA	NA	NA
Habitat provision[17]	–2	*** crops at landscape level; management of field margins and riparian areas	*** topography, soil properties	NA	0	1 [negative impact on field margins and riparian areas]	–	1–2 [increased habitat heterogeneity; extent of change dependent on forest type and forest management[18]]	+
Soil formation and structure	1	* organic matter, bacteria, fungi; roots	** climate, parent material, erosion	0–3 [dependent on soil type and pre-parasite conditions]	–/+	0–3 [dependent on soil type and pre-drought conditions]	–/+[19]	0 [dependent on litter quality[20] and quantity, humification and sequestration in biogenic structures]	+

(continued)

Appendix Table 5.A1. (*continued*)

Goods and Services	Unmanaged (Arable Tilled)[1]			Perturbations						
				Invasive Species: Root Parasites[2]		Climate Change: Drought[3]		Land-Use Change: To Forest		
	Service Rank	Biotic	Abiotic	Vulnerability	Net Impact	Vulnerability	Net Impact	Vulnerability	Net Impact	
Nutrient cycling	3	** microbial activities including mycorrhizae micro-food web control	* climate, soil properties P-cycling	0–1 [dependent on whether biotic factors, e.g., microbial activities, micro-food web slowed]	– / +	2 [biological activity decreased[21]; possible further reduction in NPP through salinization]	–	0 [change to nutrient conserving mechanisms]	0	
Biodiversity	–3	* roots, biogenic structures	** climate, soil resource quality microhabitats	0–1 [dependent on impact on rhizosphere biota]	– / +	0 [dependent on abiotic drivers, resource quality, microhabitats, roots]	–[22]	0 [dependent on forest type, litter quality and quantity, increased biotic processes and invertebrate engineers]	+	

[1] Other arable systems—for example, no-till, ridge-till—are not considered, but subject to similar vulnerability; magnitude may vary.

[2] Considers general case of root parasite invasion, for example, root knot nematode (*Meloidogyne* spp.).

[3] *Drought* is defined as 20 percent decrease in summer precipitation, based on an average for a 10-year period. This is in contrast to episodic drought, for example, in semi-dry African savanna.

[4] Drought can have a significant economic impact on agriculture. For example, the cumulative damage from the four drought years (1996, 1998, 1999, and 2000) to the agricultural industry in Texas is US$5.515 billion since 1996. (Myers 2002; see also http://www.txwin.net/dpc/dpc%20biennial%20report.pdf)

[5] Plant varieties are bred for resistance to drought, but success is variable. See (http://www.hort.purdue.edu/newcrop/proceedings1999/v4-060.html).

[6] Hunt et al. 1989.

7 Bingham 2001.

8 Refers to runoff to streams.

9 There are many levels of drought and types of managed arable soils, thus vulnerability and net impact is highly context dependent. For example, in the prairies under drought conditions, farmers allow fields to go fallow to store water. However, soil is left bare, and if drought continues, soil erosion by wind and the loose stability of the soil surface are negative consequences.

10 Response by farmers to drought can include switching the crop planted.

11 Refers to biological control of diseases and pests of crop plants.

12 Refers to genetically modified crops.

13 Can have change in types of pests under extensive drought. For example, grasshopper and locust populations can be extensive, some nematodes can thrive, but generally caterpillar populations are reduced. Furthermore, there may be a possible exacerbation of plant-pest interaction with drought: for example, plants previously weakened by drought may be more susceptible to pests.

14 Service of recreational land use is very culturally dependent and dependent on type of forest planted: for example, landscape with a monoculture of tree species may not be an improvement on a landscape of crop monocultures.

15 There is a gradient in microbial activity, decomposition, and organic matter buildup along a drought gradient. Whether a drought starts in a moist situation or in a dry situation will influence services such as carbon sequestration (www.ornl.gov/carbon_sequestration/).

16 Service of carbon sequestration is very dependent on forest type; for example, although pine forests have the same aboveground productivity as deciduous forests, they have lower carbon sequestration.

17 Habitat provision for high visibility and/or threatened species.

18 Habitat provisioning may not increase significantly if land-use change is from annual monoculture to perennial monoculture.

19 Effects of drought on soil formation and structure are very soil texture dependent. For example, drought has a major effect in sandy soils, but effects are less dramatic in clay soils.

20 Service of soil formation is very dependent on forest type: for example, soil formation under pine can lead to development of podsol.

21 Buljovic & Engels 2001.

22 With severe drought, earthworms will be lost from soil, but numbers of termites and ants may increase. In general, biodiversity in cultivated soil is only marginally affected by drought, as most species under arable conditions are resistant and resilient.

Appendix Table 5.A2. Vulnerability of ecosystem goods and services in unmanaged grassland ecosystems provided by the soil biota to three agents of global change: invasive species, climate change, and land-use change. Scoring system used is as for Table 5.A1.

| Goods and Services | Unmanaged (Grassland)[1] | | | Perturbations | | | | | |
| | Service Rank | Biotic | Abiotic | Invasive Species: Plant Species[2] | | Climate Change: Drought[3] | | Land-Use Change: To Arable Land | |
				Vulnerability	Net Impact	Vulnerability	Net Impact	Vulnerability	Net Impact
Food production[4]	2	*** decomposition, nutrient cycling, bioturbation, resistance to pests and diseases	** soil type, topography, fire, precipitation	1–3 [unpalatable species, indirect and subtle effects on soil food web]	–5	2–3 [reduced NPP[6]]	–	3 [change from fungal to bacterial-based food web[6], reduced earthworm abundance and mycorrhizal infection, build up of pests and diseases]	– / 0 / +
Water quality and quantity, Flood, erosion control[7]	2	*** retention of N in biomass, physical stabilization, soil aggregate formation increasing infiltration rates, interception of runoff, moisture retention by organic matter	** soil type, topography, fire, precipitation	1 [transpirational losses from deep-rooted shrubs and trees[13]]	–5	3 [increased particulate matter and nutrient load in surface runoff and erosion[8], flush of biological activity following re-wetting events, increased plant uptake and decreased retention]	–	3 [nitrification, DON, phosphate, DOP, increased surface runoff and erosion[8] resulting from decreased plant cover and destabilized soil surface]	–

Animal fiber Wool, leather	2	*** decomposition, nutrient cycling, bioturbation, resistance to pests and diseases	*** soil type, topography, fire, precipitation	1–3 [unpalatable species, indirect and subtle effects on soil food web]	−	2–3[10] [Reduced NPP]	−	2–3[10] [decline in grazing animals]	− / 0 / +[11]
Plant fiber	1	** plant breeding, roots and residues	*** climate, tillage, soil type, fertilizer	3 [can lead to destruction of crop]	−	3 [abiotic controls, especially soil properties and fallow practices, most important; reduced NPP]	−	3 [litter system developed; annual roots replaced by perennial; increased role for soil food web; change from fungal- to bacterial-based food web[6]]	+ / 0 / −
Recreation and other uses	2	*** decomposition, nutrient cycling[12]	** soil type, topography, fire, precipitation	1 [altering system structure, decreased native animal grazing]	−	1 [potential loss of landscape features and loss of some fauna as a result of changes in vegetation]	−	1 [altering system structure, decreased native animal grazing[14]]	− / +[13]
Carbon sequestration	2	*** organic matter accumulation, CaCO3 deposition	** complexing organic matter, texture, fire, CaCO3 deposition	1 [accumulation of woody material and recalcitrant litter, decreased animal grazing]	+	2–3[9] [reduced NPP]	−[15]	3 [enhanced decomposition and microbial access to C within soil aggregates]	−[15]
Trace gases and atmospheric regulation	2	*** maintenance of C and N balances	** texture, precipitation, pH	0	NA	1 [modified changes in vegetation]	−	3 [increase in NO due to increased nitrification and denitrification rates]	−

(continued)

Appendix Table 5.A2. (*continued*)

	Unmanaged (Grassland)[1]			Perturbations					
				Invasive Species: Plant Species[2]		Climate Change: Drought[3]		Land-Use Change: To Arable Land	
Goods and Services	Service Rank	Biotic	Abiotic	Vulnerability	Net Impact	Vulnerability	Net Impact	Vulnerability	Net Impact
Habitat provision[16]	3	*** decomposition, nutrient cycling, bioturbation, resistance to pests and diseases, biota providing food resources and habitat	* soil type, topography, fire, precipitation	1–3 [unpalatable species, indirect and subtle effects on soil food web, altering habitat structure]	– / 0 / +[17]	1[18] [change in vegetation structure, shift in taxa due to drought leading to increased small mammals and predatory bird populations]	+ / –[19]	3 [shift to a relatively homogeneous habitat, change from fungal to bacterial-based food web, removal of native plants]	—
Waste disposal/ bioremediation	1	*** decomposition of brewery waste and sewage sludge	* compounds binding to humic material and clays	1 [unpalatable species, indirect and subtle effects on soil food web altering of microflora to degrade organic compounds]	– / 0 / +	2 [increased heavy metal concentration, decreased decomposition rates]	—	3 [increased accumulation of heavy metals from sewage sludge applications]	—

Biological control[20]	1	*** native soil pathogens and predators may inhibit the rapid spread of exotic and ruderal plant species	* soil type	1 [individual plant species may influence soil communities, activity, and/or function[21]]	–	1 [changes in climate may positively or negatively impact native soil community]	–/+	1 [possible decrease in soil pathogen richness]	–
Soil structure	3	** organic matter, bacteria, fungi, roots, invertebrate fecal pellets, invertebrate[22] engineering[23]	** climate, parent material, topography, erosion	0–1 [may exert strong impacts on soil biota]	–/+[24]	0–1[25]	–	3 [altered soil structure due to plowing, compaction, increased salinity, decreased fungal biomass, simplified soil food web]	–
Nutrient cycling	3	*** organic matter, microbial activity, microbial root symbionts, invertebrate microbial grazers	*** climate, soil texture, topography, pH, dry and wet deposition of nitrogen and sulfur	0–2 [may influence soil biota or be a novel nitrogen fixer]	–/+[26]	2 [decreased plant and soil biota activity]	–	3 [shift from a fungal to bacterial-based food web, simplified soil food web, additions of inorganic fertilizers]	–/+[27]
Biodiversity	3	** the presence of plants can influence the community composition of soil biota[28]; plant species can influence soil microbial physiological attributes[29]	** climate, physical and chemical soil heterogeneity	0–2 [altering soil physical and/ or chemical properties, facilitate persistence of exotic invertebrates[30]]	–	1[31] [decreased plant inputs into soil, low soil moisture]	–	3 [simplified food web, decreased soil structure, increased soil salinity, limited plant input into soil due to harvesting, relatively homogeneous plant community, biocides and GMOs influencing non-target soil taxa[32]]	–

(continued)

Appendix Table 5.A2. (continued)

| Goods and Services | Unmanaged (Grassland)[1] | | | Perturbations | | | | | |
| | Service Rank | Biotic | Abiotic | Invasive Species: Plant Species[2] | | Climate Change: Drought[3] | | Land-Use Change: To Arable Land | |
				Vulnerability	Net Impact	Vulnerability	Net Impact	Vulnerability	Net Impact
Fuel	0	NA	NA	NA	NA	NA	NA	NA	NA
Biochemicals/ medicines	1	** plant compounds and microbial enzymes	** NA	1 [possible loss of native plant species]	– / +[33]	1 [changes in microbial community]	– / +[34]	3 [change from fungal- to bacterial-based food web, removal of native plants]	–

[1] Lightly managed free-range grasslands (i.e., including supplemental winter livestock feed, but excluding fertilizer or pesticide inputs).

[2] Considers general case of a plant species invasion.

[3] Drought is defined here as a 20 percent decrease in annual precipitation, based on an average for a 10-year period.

[4] Animal products are considered here.

[5] Hirsch & Leisch 1998; Jacobs & Sheley 1998.

[6] Hunt et al. 1989.

[7] Refers to runoff to streams. In dryland systems, potential evapotranspiration generally exceeds precipitation, so these services may be less significant.

[8] Branson et al. 1981.

[9] Range depends upon organic matter and moisture retention. Therefore this good and/or service will be more susceptible in more xeric, low–organic matter grasslands.

[10] Range depends upon whether the arable land supports winter grazing by cattle (e.g., on winter wheat fields) and if fiber crops are planted (e.g., cotton).

[11] Direction of net impact will probably be negative, but could be positive if fiber production in arable land is higher (e.g., from the production of cotton) than wool or leather production of the original grassland.

[12] Hikers, bicyclists, birders, hunters, etc. are often attracted to a site because they find it visually pleasing and/or because the site serves as habitat for charismatic and/or edible animal taxa (e.g., birds and deer). In part, it is the feedbacks between the soil and plant systems that result in the visually appealing structure (from a human perspective) and produces habitat and food for animals of interest to the public.

[13] Some charismatic animals may increase in arable lands (e.g., some game bird species).

[14] Creating a more structurally homogeneous landscape will limit the number of charismatic animals found at the site and thus will decrease public interest in using the site for recreational purposes.

[15] Burke et al. 1989.

[16] Habitat provision for high visibility and endangered/threatened species.

[17] Direction of net impact will depend upon whether or not native taxa can utilize the invasive plant species for food or habitat.

[18] Assumes that the taxa important in habitat provision are adapted to the ecosystem and the ecosystem is normally subject to drought.

[19] Impact could be positive or negative depending upon the response of important taxa to drought. Some taxa may be favored in drought conditions.

[20] Refers to biological control of parasites and diseases of native species.

[21] Wardle et al. 1999; Kourtev et al. 2002.

[22] Rusek 1985.

[23] Refers to termites and earthworms Lavelle et al. 1997.

[24] Impact could be negative or positive depending on if the invasive plant species inhibits or stimulates soil biota via root exudates, above- and/or belowground litter inputs, or altering soil microclimate.

[25] Assumes biota are adapted to periodic drought conditions.

[26] A negative impact suggests that the rate of cycling of a given nutrient is slowed with the addition of an invasive plant species while a positive impact suggests an increase in the rate of cycling. Whether or not humans desire an increase or decrease in nutrient cycling rates is probably context dependent.

[27] The net positive impact suggests an increase in the rate of nutrient cycling that may benefit crop plants. However, this increase can also lead to increased nitrogen and carbon losses that may contribute to greenhouse gas emissions.

[28] Wardle et al. 1999.

[29] Westover et al. 1997. Changes in microbial physiological traits may be suggestive of a change in microbial community structure.

[30] Kourtev et al. 1999. Facilitating the persistence of nonnative invertebrates may decrease the biodiversity of native invertebrates.

[31] Assumes biota are adapted to periodic drought conditions and will probably reach pre-drought biodiversity relatively quickly following the end of the drought.

[32] GMOs are genetically modified organisms. There is evidence that the insecticidal toxin from the bacterium *Bacillus thuringiensis*, encoded into some crop plants, is released from the engineered plants into the soil environment (Saxena & Stotzky 2000).

[33] Direction of net impact will be affected by any biochemical/medicinal products that could be obtained from the invasive plant species.

[34] Direction of net impact will depend upon what components of the microbial community are inhibited or favored during drought conditions and whether those microbes can yield any biochemical/medicinal products.

Appendix Table 5.A3. Vulnerability of ecosystem goods and services in unmanaged forest ecosystems provided by the soil biota to three agents of global change: invasive species, climate change, and land-use change. Scoring system used is as for Table 5.A1.

| Goods and Services | Unmanaged (Forest)[1] | | | Perturbations | | | | | |
| | Service Rank | Biotic | Abiotic | Invasive Species: Exotic Earthworms | | Climate Change: Drought[2] | | Land-Use Change: Deforestation | |
				Vulnerability	Net Impact	Vulnerability	Net Impact	Vulnerability	Net Impact
Food production	1	*** fungi, fruits for animals	* pH	1 [altered fungal diversity, increased fungal dominance[4]]	–	1 [reduced fruiting body abundance[5], changes in phenology/synchrony]	–	1 [loss of mushrooms, fruits, hunting, etc.[6]]	–
Water quality and quantity Flood, erosion control	3	** plant species composition, soil organisms, bioturbators	*** topography, soil texture, porosity	1 [increased soil erosion, nitrate leaching[7]]	–	3 [reduced water yield, solute concentrations increased]	–	3 [increased decomposition, increased nitrification, acidification, increased water yield[8]]	–
Fiber (wood) production	3	* pathogens, parasites, decomposers, N_2 fixers	*** soil factors affecting NPP	0 [increased NPP[9]]	+	3 [reduced NPP[10]]	–	3 [absence of trees]	–
Fuel[11]	3	* pathogens, parasites, decomposers, N_2 fixers	*** soil factors affecting NPP	0 [increased NPP]	+	3 [reduced NPP]	–	3 [absence of trees]	–
Biochemicals/medicines[12]	1	** microbial diversity	*	0	+	0	–	0	–

Service	Score	Organisms	Factors						
Habitat provision[13]	3	*** decomposers, ecosystem engineers	*** humus type, pH, soil fertility, parent material, topography	0	0	1 [changes in understory vegetation[14]]	−	2 [changes in habitat types[15]]	—
Waste disposal/bioremediation	1[16]	* decomposers, ecosystem engineers	* forest type, pH, soil fertility, parent material, topography	0[17]	0	2 [reduced decomposition rates, increased solute concentrations[18]]	+	1 [higher throughput of water[19]]	—
Biological control[20]	1	* insect taxa, mycorryhizal and fungal pathogens	** water-logging, pH, soil texture and structure	0 [modification of plant growth]	+	1 [unpredictable disease interactions[21]]	+	1 [stumps may be important disease sites[21]]	—
Recreation and other uses	3	** ants and insects for wildfowl, game	*** structure of the vegetation[22]	0 [increased resources for birds and other animals, fishing]	−	3 [increased fire risk, drier soils may ease access]	−	2 [tree loss, reduced hiking, negative visual impact]	—
Carbon Sequestration deciduous forest	3	* litter quality, worms, roots	*** soil texture	1 [rapid litter loss]	−	2 [reduced NPP]	−	3 [increased decomposition, major C loss]	—
Trace gases and atmospheric regulation	3	*** nitrifiers, methane oxidizers, denitrifiers	*** tree species, pH, soil texture, aeration[23]	1 [increased N_2O production[24]]	−	1 [decreased N_2O production, increased CH_4 oxidation[25]]	+	2 [increased N_2O production, decreased CH_4 oxidation[25]]	—
Soil Structure	3	*** earthworms, roots, arthropods, bacteria, fungi	** climate, parent material, erosion	0–3 [dependent on aspects of soil structure considered[26]]	+/−	1 [some reduced activity and leaching]	−	3 [compaction, erosion, soil community shifts]	—

(continued)

Appendix Table 5.A3. (continued)

	Unmanaged (Forest)[1]			Perturbations					
				Invasive Species: Exotic Earthworms		Climate Change: Drought[2]		Land-Use Change: Deforestation	
Goods and Services	Service Rank	Biotic	Abiotic	Vulnerability	Net Impact	Vulnerability	Net Impact	Vulnerability	Net Impact
Nutrient cycling	3	*** microbe-faunal interactions	* climate, soil properties, topography, P cycling	3 [accelerated cycling of nutrients]	+/-	2 [biological activity decreased]	–	3 [higher rates of decomposition and leaching from system]	–
Biodiversity	2	***, roots, substrate for soil organisms	* climate, soil resource quality, microhabitats	3 [displacement/ extirpation of native soil organisms]	+/-	0 [dependent on abiotic drivers, resource quality, microhabitats, roots]	–	3 [community shifts to opportunistic species, increased dominance]	–

[1] Entries are for a typical temperate deciduous forest, principally managed for timber yield.

[2] Drought is defined as 20 percent reduction in summer precipitation, based on an average for 10-year period. The important impacts of fire have not been considered, but risk of fire would increase dramatically under drought conditions.

[3] Deforestation is assumed to be conventional felling, involving timber removal but non-removal of felling debris.

[4] Introduction of earthworms into forest soils has been shown to increase the diversity of fungi in burrow walls (Tiunov & Dobrovolskaya 2002), but can also have an effect on mycorrhizal colonization and, hence, potentially fruiting bodies (Lawrence et al. 2003).

[5] Dry years are typically associated with strongly negative effects on fungal fruiting bodies (e.g., see Wiklund et al. 1995). This importance of forests as sources of edible fungi differs greatly between regions.

[6] Forests act as habitats for a number of food species, including deer and boar. The overall impact of deforestation on these foods depends on the subsequent use of the land after deforestation, but a felled forest will generally act as a poorer service provider in this respect.

[7] Experimental evidence suggests increased N mineralization and nitrification after earthworm introduction (Robinson et al. 1992). This may have local impacts on soil solution chemistry but is unlikely to have a major impact on stream chemistry.

[8] Classic studies (e.g., Hubbard Brook) have clearly demonstrated the impacts of deforestation on forest streamwater chemistry. The effects are largely due to changes in deposition and hydrology, due in turn to canopy removal and to the dramatic switches in physical status of soil that impacts decomposition rates.

[9] In the laboratory, increased growth of tree seedlings has been demonstrated as a result of earthworm addition (Haimi & Einbork 1992).

[10] Decreased NPP with drought, with slower growth. Timber quality may actually improve, however. Mycorrhizal fungi may play a significant role in forest drought resistance (Garbaye 2000).

[11] Only the provision of wood is considered.

[12] Current role of temperate forests as sources of biochemicals and medicines is limited in temperate regions. However, future potential and nontemperate forest use cannot be totally ignored and the loss of even one plant species may be of importance.

[13] Habitat provision for high-visibility/threatened species.

[14] Changes in understory vegetation may occur as a consequence of drought. Reductions in understory density may reduce value as a habitat.

[15] Deforestation may have a range of impacts on small and large mammal, bird, and invertebrate populations.

[16] Production forest has been evaluated as disposal sites for sewage sludge (Ferrier et al. 1996); the disposal process depends heavily on the activities of soil biota. Production forest is also a suitable land use for revegetated mine wastes.

[17] Interaction not investigated, but evidence suggests that earthworms may help break down waste. Reclamation engineers use the viability of earthworms as an index of soil condition.

[18] Shortage of water would have the direct physical effect of decreased dilution of wastes, while limitation to microbial activity could lead to decreased degradation.

[19] Increased channel bypass flow and decreased interaction time between water and soil would reduce ability to deal with certain pollutants (Bardgett et al. 2002).

[20] Refers to biological control of parasites and diseases of natural species.

[21] Strong interactions between site management and root pathogens with drought have potentially beneficial effects, whereas untreated stumps after felling may act as sources of root pathogen inoculum (Thies 2001).

[22] Tree species and planting density have a major affect on recreational use of forests, for both hikers and hunters.

[23] Trace gas consumption by forests depends on water status, species, management and, in particular, extent of fertilizer addition (Ineson et al. 1991).

[24] Associated with increased N mineralization and earthworm casts (Elliott et al. 1991).

[25] Soil water content has a major effect on methane transfers in soils, with optimum contents for methane oxidation and nitrous oxide production (Del Grosso et al. 2000). Both drought and deforestation have marked effects on these fluxes.

[26] Earthworms contribute to soil structure through bioturbation and redistribute organic materials over a large vertical range in the soil profile; however, as invasives they displace enchytraeids and microarthropods that are also critical for soil structure and development.

6

Vulnerability and Management of Ecological Services in Freshwater Systems

Paul S. Giller, Alan P. Covich, Katherine C. Ewel, Robert O. Hall, Jr., and David M. Merritt

Society obtains great benefits from properly functioning ecosystems in the form of provisioning (e.g., food), supporting (e.g., waste processing, sustained supplies of clean water), and enriching (e.g., recreation) services, all of which are provided at multiple scales and at no charge to society. Freshwater benthic ecosystems often play important and unique roles in providing many of these services (see Chapter 3, Table 3.1), but the number and magnitude of anthropogenic stressors that threaten these services is growing rapidly. Sustainable development, which "meets the needs of the present generation without compromising the ability of future generations to meet their own needs" (World Commission on Environment and Development 1987), depends on our ability to manage and maintain these ecosystems and the services they provide. In order to achieve this end, we need a better understanding of how benthic ecosystems function and are structured, as well as stronger integration of management with ecological studies. Having insulated ourselves from many natural ecosystems through technology, we often fail to appreciate the beneficial "workers" that sustain "nature's economy" upon which ecosystem services depend.

We now appreciate that degradation of freshwater sediments, which harbor the biota essential to benthic ecosystem processes, will in turn degrade water quality and a range of other services. We also understand that because of the strong linkage between freshwater ecosystems and the landscapes they drain (Giller & Malmqvist 1998), changes in land use and other activities in the catchment can contribute to such degradation. Many

We are very grateful to three anonymous referees for their constructive insights and comments. We are also grateful to Willem Goedkoop for his contributions to the discussions and ideas that initiated this chapter.

current water management practices, such as flood control, water diversion and deten-tion, channelization, and irrigation, affect the hydrological cycle at local to catchment scales. Over the past several hundred years, humans have built thousands of kilometers of diversion canals, channels, and levees to divert water for society's use. Humans have drained wetlands for urban development and agriculture, and have dammed rivers for water abstraction and the generation of hydroelectric power. Although these activities are intended to provide certain important services to the human population, they also sig-nificantly degrade many other services, the values of which become evident only when they are lost or destroyed. The examples of the demise of the Aral Sea in Central Asia (Micklin 1992) due to the diversion of inflowing freshwater streams, and the ongoing threats from paper mill effluent to the species-rich and globally unique freshwater biota of Lake Baikal in southeast Siberia, are clear cases in point. On a larger scale, climate change, an unintended consequence of human activities, also alters the hydrological cycle, threatening freshwater habitats and organisms (and hence a range of services) through-out the world (Palmer et. al. 1997; Lake et al. 2000; Wall et al. 2001). The threats of such activities on sustainable development are clear.

The various types and importance of ecosystem services in fresh waters, and the role of benthic biodiversity in the delivery of these services, are presented in Chapter 3, along with a discussion of the balance between ecological and economic values. In this chap-ter, we will briefly review the various threats to freshwater benthic ecosystems and the important benthic species that help sustain ecosystem services. We also consider the vul-nerability of these services, using a number of case studies to illustrate the cascading effects of overexploitation and the subsequent loss or degradation of other services. These case studies also illustrate how benthic organisms and the ecosystem services they perform can be used to enhance management and maintain the overall health and sus-tainability of freshwater systems.

Threats to Freshwater Systems

Threats to freshwater systems arise from a myriad of human activities, including chan-nelization, groundwater pumping, diversion, dam building, pollution, human-induced climate change, and overexploitation of natural resources (e.g., Postel & Carpenter 1997; Malmqvist & Rundle 2002). Nearly all major rivers and lakes worldwide have large human population densities associated with them or within their drainage basins, usually sited there with relatively little thought to the availability of potable water. The growth of the human population and the mismatch between population growth and provision of, and accessibility to, water resources is an imminent concern (Cohen 1995). An estimated 1.8 billion people now live under a high degree of water stress in areas with limited supplies of potable water (Vörösmarty et al. 2000). This stress may continue to rise, with a projected population living in these areas estimated to be between 2.8 billion and 3.3 billion by 2025 (Engelman & LeRoy 1993, 1995; Cohen 1995).

Stressors and impacts that force changes in freshwater ecosystems can be classified into four major types of threat (Malmqvist & Rundle 2002): (1) complete ecosystem loss or destruction, (2) physical habitat alteration, (3) water chemistry alterations, and (4) modifications of species composition. Ecosystem loss or destruction is often associated with water withdrawal from the system (e.g., in the Alps, Ward et al. 1999) resulting from rapid urbanization and/or intensification of agriculture, and the associated water demand and lowering of water tables by extraction elsewhere. There is a strong correlation between population size and water withdrawal (Gleick 2001), and irrigation dominates water demand at the global level. Habitat alteration of the freshwater system can occur from both instream activities (including channelization, damming, and draining of wetlands) and catchment-related activities (such as deforestation, poor land use, and alteration of the riparian corridor). Changes in water chemistry result from pollution due to wastewater discharge, diffuse nutrient loading from agriculture runoff, acidification from atmospheric inputs, and the introduction of endocrine disruptors (Malmqvist & Rundle 2002). Introductions of exotic species may be direct or indirect (as discussed below). Extinctions are common, often due to overexploitation of the organisms themselves, habitat destruction (or loss of habitat to invasive species replacement), the loss of functions necessary for some life stage of a particular species, or the loss of a symbiont.

We have identified 14 major threats to the six major services provided by freshwater benthic systems (Figure 6.1). Each threat can impact more than one of the services, and many of these impacts are mediated through the benthos. In reality, each threat can be subdivided into a finer series of threats. For example, hydrologic modification can have effects through a decrease in peak flow, increase in low flows, change in timing of peak flows, changes in the rate of drawdown, and/or a decrease in flow variability, and so on. Each ecosystem service can be affected by several different threats, and different stressors may act synergistically. Eutrophication can increase biotic activity and thereby enhance the effect of metal contamination (for example, the mobility of mercury). Likewise, changes in water chemistry, mechanical disturbances to a system, or changes to the characteristics of the habitat can enhance the probability of successful species invasion (Jenkins & Pimm 2003), which in turn may decrease economic success based on a highly profitable food source for humans. Changes in the competitive balance between species can also ensue. One example of this phenomenon is the replacement of the sawgrass (*Cladium jamaicense*) communities in the wetlands of the Everglades in Florida, United States, by cattail species (*Typha latifolia* and *T. domingensis*) as a result of phosphorous and nitrogen loading from agricultural runoff (Newman et al. 1998). In areas of the 600,000–ha Everglades that have the highest phosphorous enrichment, cattails dominate, but in portions of the Everglades where phosphorous remains low, sawgrass still dominates. This shift in community structure directly resulting from human-caused changes in water chemistry is due to the fact that cattails are better able to assimilate nitrogen and phosphorous and to produce biomass.

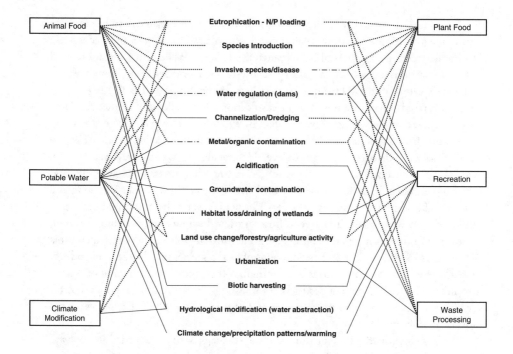

Figure 6.1. The interaction between six major ecosystem services, provided by freshwater systems, and fourteen potential threats in the freshwater domain. An explanation of the nature of the services is given in Chapter 3, Tables 3.1a–3.1e. Solid lines indicate the direct links between the major services and the various threats, and the dotted lines indicate links that may be mediated through the benthos.

The stressors described above in Figure 6.1 occur in all types of freshwater ecosystems; however, the magnitude and direction of their effects vary across ecosystems. Lakes and wetlands are susceptible to various stressors due to their slow turnover of water, their potential for accumulation of toxins and metals in their sediments, and their dependence on the quality and quantity of water inputs from inflow streams. The susceptibility of rivers and associated wetlands, on the other hand, is exacerbated by the downstream flow of water (and hence pollutants and sediments) and their longitudinal connectivity (upstream and downstream dispersal migration of many species). Almost any significant activity within a river catchment and throughout its drainage network may have the potential to exert effects for large distances upstream and downstream.

Freshwater ecosystems face different threats in different regions, depending largely on the economic activity and state of development. Water is abundant at high latitudes and in the wet tropics; however, in much of North and East Africa, Australia, and parts of North America, the availability of potable water is relatively scarce. Even in the more temperate countries with relatively high overall annual precipitation, major concentrations of population are often located in areas of lowest rainfall (such as Dublin and London), creating local water deficits that require large-scale engineering projects for water storage and/or transfer, as well as water regulation activities to overcome. Roughly 40 percent of the world's population that live in 80 dry, or partially dry, countries face serious periodic droughts (Cohen 1995); these pressures on water resources will be more pronounced in Africa and South America by 2025 (Vörösmarty et al. 2000). Plans to redirect water from uninhabited areas to population centers will create additional problems. Lakes in the developed world are threatened by eutrophication and lowered water tables due to groundwater abstraction, while in the undeveloped world, overexploitation of fish and invasion from exotic plants (e.g., the water hyacinth *Eichhornia crassipes*) are more problematic. Destruction of running water habitats is extensive in much of the developed world (because of flood control, drainage, clearing channels for transportation and transport of timber, and dredging), as well as in the developing world (largely due to dam construction and mining; see Covich et al., Chapter 3).

Waste disposal poses significant threats to many systems, as treated and untreated domestic and industrial waste leads to significant levels of eutrophication and to metal and other chemical contamination. Sedimentation and nonpoint source pollution result from changing land use such as deforestation, overgrazing, and intensification of agriculture. The degradation of riparian zones that often accompanies such intensification (as in the Netherlands, for example) also changes benthic ecosystem functions dramatically (Gregory et al. 1991). Even atmospheric pollution impacts aquatic ecosystems, as evidenced by acidification of freshwater systems throughout northern Europe, the northeastern United States, and Canada (Stoddard et al. 1999).

Anthropogenic threats and influences alter the balance of natural regulatory factors in freshwater systems such as energy supply and flow, organic and inorganic matter transport, hydrologic regimes, hydrologic and biogeochemical cycles, and water chemistry (Malmqvist & Rundle 2002). These anthropogenic factors change the structure of freshwater sediment, alter temperature regimes, and cause other environmental conditions to change beyond the normal levels of variation and extremes. Such changes will clearly impact species unless they possess certain traits that confer resistance or resilience to the environmental change.

Interaction of Threats and Ecosystem Disservices

There is frequently a trade-off between ecological and economic values associated with ecosystem services (see Covich et al., Chapter 3). In this context, the interconnections between services and threats provide an introduction to the concept of disservices. Exploiting one service can negatively affect, or in extreme cases completely eliminate,

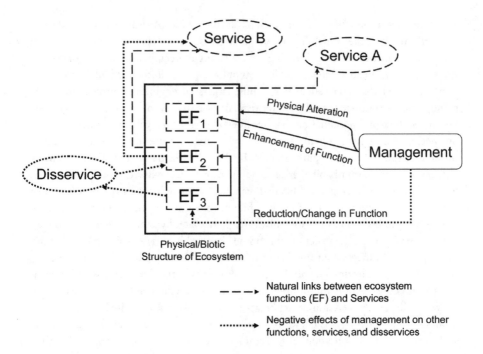

Figure 6.2. Conceptual diagram illustrating a disservice for humans, through the potential negative effect of implementation of management on an ecosystem function (EF_1) to enhance an ecosystem service (Service A). In this example, the physical alteration of the structure of the ecosystem through the management regime also leads to a reduction or change in function of EF_3. This in turn can act as a disservice to EF_2, hence causing a negative impact on Service B.

another service. The potential for disservice is exacerbated when society introduces a management process to enhance one particular ecosystem function, and hence one particular service, that unintentionally leads to a reduction or change in another ecosystem function. The net effect is to degrade a second, non-target, ecosystem service (Figure 6.2).

For example, regulating a river to obtain hydroelectric power has negative consequences on ecosystem services that depend on natural hydrologic regimes and free-flowing water. Natural rivers and streams are heterogeneous systems with flow patterns that vary over time, often in predictable patterns (Poff & Ward 1989). However, in regulated rivers, the purpose of management is to adjust this variation; the reduced seasonal and increased daily dynamics of flow in regulated rivers instead reflect the periods of electricity demand. Reduced spring floods and elevated winter discharge, and even high daytime/weekday flow and low nighttime/weekend flow often result (Malmqvist & Englund 1996). In addition to changing flow regimes, habitat loss, fragmentation, and (in heavily regulated systems) a general change from lotic (running water) to lentic (standing water), or even to dewatered conditions, can occur along some stretches of river. These flow alterations change river habitats and their biota, including sedimentary biota, fundamentally. For example, species richness and abundance of benthic macroinvertebrates are impacted (Ward & Garcia de Jalon 1991) which, given the clear role of such invertebrates in delivery of ecosystem services (see Covich et al., Chapter 3), will lead to degradation of such services (e.g., breakdown of wastes, support of fisheries, and maintenance of high water quality). Effects on migrations and stocks of anadromous salmonids, in turn, alter various services such as recreation. This is of significant concern, for instance, in Sweden (Malmqvist & Englund 1996; Jansson et al. 2000), where over 70 percent of the rivers are regulated, and also in the northern third of the Earth, where a majority of rivers are heavily influenced by river regulation (Dynesius & Nilsson 1994). Worldwide, the number of "large" dams (greater than 15 m) has increased approximately eightfold over the last 50 years and the total number of dams has risen to nearly 800,000 (Postel & Carpenter 1997; Rosenberg et al. 2000).

The introduction of exotic species further illustrates the concept of the interaction of threats and ecosystem disservices. Exotic species introductions create less visible but ecologically and sometimes economically important problems. Introductions of non-native fish to improve fisheries for sport or to provide protein for human consumption have led to the wholesale collapse of local biological communities and aquatic food webs. The negative impacts on ecosystem functions that rely on intact food webs, such as decomposition, biogeochemical cycling, and overall productivity, will likely affect water quality, other fisheries, and hence a number of other services. The introduction of a predatory cichlid into Gatun Lake in the Panama Canal (Paine & Zaret 1973) and the introduction of Nile Perch into the African rift valley lakes (Kitchell et al. 1997) are two cautionary cases. The zebra mussel (*Dreissena polymorpha*) is perhaps the most famous and pervasive invading freshwater exotic. Originally native to the Black and Caspian Seas, it spread throughout European waters during the 19th century, but it has only recently reached Ireland (Minchin et al. 2003). Discovered in 1988 in Lake St. Clair in the Great Lakes of the United States, it rapidly spread throughout the Great Lakes and Mississippi river basin (Ludyanskiy et al. 1993). There is a range of abiotic, biotic, direct, and indirect impacts of a zebra mussel introduction (MacIsaac 1996), but one of the greatest is the impact of biofouling (undesirable accumulation on artificial surfaces) associated with the settlement of the mussels. Water intake structures for municipal, industrial, and hydroelectric plants are extremely vulnerable to fouling if they draw intake water from an infested water body. Reaching densities of up to 334,000 mussels per square meter, the exotic mussels have significant ecological effects such as competitive exclusion of native unionid mussel species (Ludyanskiy et al. 1993; Fitzsimons et al. 1995) and removal of phytoplankton, which results in the disruption of food webs (Karatayev et al. 1997). Theoretical estimates suggest that the native mussel population in Lake Erie can filter up to 14 times the entire lake and remove phytoplankton equivalent to 25 percent of primary production each day (Bronmark & Hansson 1998). On the other hand, this trait can be utilized in management of eutrophication problems as seen in the Netherlands. By efficiently filtering phytoplankton, these mussels are able to alter nutrient cycling patterns, transfer carbon and nutrients from the pelagic to the benthic zone through the build up of mussel biomass, reduce the concentration of suspended solids, and hence improve water clarity and increase macrophyte growth.

Given the various threats to ecosystem functioning in freshwater systems, one might ask why the problem is not worse. After all, large population centers are still supported by surface waters that continue to provide a number of ecosystem services. The answer lies in part with the level of technology and infrastructure we are able to bring to bear, such as wastewater processing, but also in the ability of the freshwater ecosystems and biological communities either to cope with a certain level of disturbance or to recover rapidly and restore function—that is, an innate resistance and/or resilience of the ecosystem. The question remains: What role do benthic biota play in reducing the impact of a threat? Specifically, how does benthic biodiversity influence ecosystem processes in freshwater systems and are there key taxa involved, the loss of which will have devastating consequences on ecosystem services? Overall, key taxa do influence ecosystem functioning; the presence of certain species can substantially influence the system (e.g., on decomposition) (Giller et al. 2004). Benthic studies, however, have generally focused on relatively low levels of species richness, and experimental studies have been at local scales (Covich et al. 1999; Covich et al. 2004). More research is needed to provide a comprehensive understanding of the various roles played by different benthic species and functional groups of species.

Management and Ecosystem Services

In this final section we present five specific case studies that illustrate, in detail, the interactions between threats and the impact on ecosystem services. We also provide examples of how some benthic ecosystem services themselves have been used to enhance management and contribute to the maintenance of the health of the freshwater ecosystems. The case studies differ in geographical location, type of freshwater system, and nature of the services and threats. The first case study highlights the Rhine and Meuse, major European rivers that have been subject to centuries of human interference leading to hydrological modifications and pollution and clear conflicts between the various ecosystem services they offer. Efforts at restoration management are reversing some of the more dramatic ecological changes. The second case study describes the Pantanal region, a huge natural wetland complex in South America that undergoes massive seasonal changes. Unlike the Rhine and Meuse rivers example, it is in a relatively undeveloped area. It offers a wide range of services to the native population and is a habitat to considerable biodiversity, yet the Pantanal is increasingly threatened from growing agriculture and mining. Lake Mendota, the third case study, provides a recreational service that was threatened by eutrophication. Here, management based on the ecological concept of the trophic cascade has been applied with some success. Like the Pantanal, the Everglades (the fourth case study) is a large wetland that provides a range of ecosystem services. Agricultural pollutants, including heavy metals, have compromised these services, illustrating the disservice phenomenon across terrestrial-aquatic boundaries. The final case study of the Catskill Mountains watershed shows the scale of watershed management that is needed to sustain water quality and provide potable water for New York City. This example illustrates how landscape management can obviate the need to replace natural ecosystem services with artificial technological processes, thus providing significant economic and ecological value.

Repairing Years of Abuse: The Impact of Management for Transport and Waste Disposal in the Lower Rhine and Meuse Rivers (Europe)

The Rivers Rhine and Meuse have served as vital European transport arteries for centuries, as well as sites for urban and industrial development and water resources. Thus, the two rivers are of considerable economic importance but have been subject to substantial anthropogenically derived changes over time. The River Rhine, a combined glacier-rainfall river, originates in Switzerland (2,200 m above sea level) and flows over 1,250 km through France, Germany, and the Netherlands with a drainage area of 185,000 km^2. In the Netherlands, it divides into three branches: Waal (65 percent of discharge), Lek (21 percent discharge), and Ijssel (14 percent discharge). The River Meuse is fed by rainwater, originates in France (410 m above sea level), and flows over 890 km through Belgium and the Netherlands, with a drainage area of 33,000 km^2. Both rivers flow into a lowland area where they form a river delta before entering the North Sea (van den Brink 1994). The earliest documented human influence on these rivers occurred in the Roman era and involved the construction of canals to regulate discharge into the Dutch Rhine tributaries (van Urk & Smit 1989) and embankments started in the Middle Ages. In the 18th century, the Rhine River floodplain was still tens of kilometers wide, and the river meandered and supported an extensive riparian forest. Large-scale river regulation began in the 19th century, with construction of dams and sluices for sea flood protection, dams for river regulation, and groins (breakwater boulder piles extending laterally into the river), weirs, and dykes to facilitate shipping. These changes impeded natural meandering and formation of side channels, cutoff channels, and oxbow lakes; consequently, the floodplain shrank dramatically (van den Brink 1994). Since then, nearly all the floodplain forests have been cut, the riparian forests have been largely removed, and the existing riparian areas have been degraded. The significant loss to the aesthetic and recreational services is self-evident.

The water quality of the main channels of both rivers has changed considerably since measurements began in the early 1900s, with increased levels of nutrients (nitrate and phosphate), salts (chloride, sodium, and sulphate), and heavy metals (cadmium, mercury, lead, and zinc) (van de Weijden & Middleburg 1989). In addition, increasing levels of organic micropollutants such as polychlorinated biphenyls (PCBs), para-aminohippuric acid (PAHs), insecticides, and herbicides have contaminated the sediments. The lower sections of the two rivers accumulate inputs from several countries upstream and are the most polluted. In the 1960s–1970s, oxygen levels were extremely low, which affected the abstraction and provision of quality drinking water. More recently, construction of sewage treatment plants has improved the Rhine, although the Meuse still suffers from low oxygen, particularly in summer (van den Brink 1994). As a result of thermal pollution from power plants and industries, water temperature in the lower Rhine and Meuse has risen by 2 to 4° C since 1900.

Not surprisingly, there have been dramatic changes in the biotic communities of the rivers. Plankton biomass in the river channels has increased, and is now dominated by a few ubiquitous centric diatoms and green algae (Admiraal et al. 1993). These add a considerable economic cost on filtration of the abstracted water. At present, the waters of the lower Rhine are dominated by sodium chloride instead of calcium bicarbonate (chloride levels increasing from < 20 mg/l in 1874 to > 200 mg/l in 1985; van den Brink et al 1990), which, together with the increased temperature, has created an environment that permitted the invasion of several exotic brackish-water and eurythermic macro-invertebrate species. These include exotic species introduced from North America and Eastern Asia and others that have immigrated from the Mediterranean and Ponte Caspian areas. One species is the benthic filter-feeding amphipod crustacean *Corophium curvispinum*, an invader originally from the southern Ponte Caspian area, which has expanded its range since 1900 from the rivers entering the Caspian and Black Seas via canals and rivers to western Europe, probably aided by shipping. It was first documented in the middle then lower Rhine in 1987, and within a couple of years increased explosively to become the most abundant species in the Rhine system. This species also reached the Belgian part of the River Meuse in 1981 and the Dutch part by the end of the 1980s. This invader has had a significant impact on the Rhine ecosystem (Neumann 2002). Its high fecundity, short generation time, and small size have led to massive densities (rising from $2/m^2$ in 1987 to $200,000/m^2$ in 1991 on stones of groins in the lower Rhine, van den Brink et al. 1993a), increased filter-feeding activity, and competition for food and space with other species, including other exotic invaders such as the amphipod crustacean, *Gammarus tigrinus,* and the zebra mussel, *Dreissena polymorpha. Gammarus tigrinus* invaded the lower Rhine in 1983, reducing abundance of the native amphipod *G. pulex. Dreissena* spp. invaded Europe from the Black Sea and Caspian Sea over two centuries ago, before the Industrial Revolution, but disappeared from the lower Rhine in 1960s due to the poor water quality and high levels of cadmium. Reductions in cadmium levels lead to *Dreissena polymorrpha's* re-establishment in 1975 and its subsequent rapid population increase. However, the species has now dramatically declined since 1987 due to competition for space with the nonnative *Corophium* (van der Velde et al. 1994). Meanwhile, a number of native brackish-water crustaceans such as the benthic amphipod *Gammarus zaddachi* range more than 100 km upstream of their original distribution boundary, as a result of increased river salinity (van den Brink et al. 1990, 1993b).

The rivers' benthic community is now largely a pollution-tolerant one, with typical pollution-sensitive aquatic insects (ephemeropteran, trichopteran, and plecopteran species) having disappeared prior to 1940. Because the latter two families are involved in detrital breakdown, the decomposition process was likely affected. Species richness of macroinvertebrates declined from 83 around 1900 to 40 in 1987. The fish community was completely dominated by cyprinids (particularly roach) in the 1970s, and anadromous and rheophilous species declined or disappeared altogether (van der Velde et al. 1990). For example, the salmon (*Salmo salar*), was overexploited and became extinct despite large-scale restocking attempts, thus negatively impacting recreational services provided by the rivers. In addition, changes in the macroinvertebrate communities and the invasion of exotic species led to changes in the river food web structure and the diet of the major predatory fish (Kelleher et al. 1998).

Over the past two decades, various restoration management measures were implemented that began to reverse some of these impacts (Jungwirth et al. 2002). The Rhine Action Plan established by the Dutch government involved all the countries bordering the river and implemented various measures to restore water quality and habitat structure. Discharge of raw sewage and industrial wastes has decreased. The much-publicized Sandoz incident in 1986 (Lelek & Kohler 1990; Mason 1996), which involved huge inputs of insecticides following a major fire in a chemical plant in Switzerland, led to the closure of water diversion plants along the river and other controls and restrictions. Despite such setbacks, some evidence of success has been seen in the rediscovery of several benthic riverine species in the lower Rhine, including the net-spinning caddis fly *Hydropsyche conturbernalis*, the water bug *Aphelocheirus aestivalis*, the damselfly *Calopteryx slendens*, and the freshwater mussels *Anadonta anatina* and *Unio pictorum* (van den Brink et al. 1990). The number of fish species has increased, rising from a low of 12 in 1971 to 25 in 1987 (van der Velde et al. 1990). Further reduction of the pollution loads in the entire drainage basin has focused on nutrients, heavy metals, and organic micropollutants (such as PCBs and PAHs). Restoration of wetland vegetation, floodplain lake water quality, and, in particular, connections between the main channel, floodplain lakes, and side channels were suggested as being particularly important from a biodiversity perspective (van den Brink 1994). Indeed, the creation of permanently flowing secondary channels on the Rhine floodplain in 1994 showed that within five years these artificial secondary channels function well as an appropriate habitat for riverine species, including the more demanding rheophilic species (those that prefer to live in running water), and have thus contributed to the ecological value of the river (Simons et al. 2001). Jungwirth et al. (2002) give a number of other examples of similar river and floodplain restoration projects.

Wetland Protection to Preserve Biodiversity and to Enhance Food Production and Recreation: The Pantanal of South America

This enormous tropical wetland, the Pantanal of South America, is approximately the size of the state of Florida and is the fourth largest complex of wetland ecosystems in the world (Keddy 2000). Its basin covers approximately 138,000 km^2 in Brazil and 100,000 km^2 in Bolivia and Paraguay. It consists of numerous streams, lakes, and seasonally flooded swamps. The basin receives inflows from several large rivers (e.g., Rio Paraguay, Rio Petras, Rio Cuiaba) that flow southward to join the Rio Parana and then to become the Rio Plate in Argentina (see Por 1995 for further details). The river water that supplies it is primarily "clearwater" (*sensu* Sioli 1984), with little suspended material under pristine conditions. A unique feature of the Pantanal is that it experiences substantial changes in the area that is under water between wet and dry seasons each year, with as little as 10 percent of the area inundated during the dry season and as much as 70 percent during the four-to-six-month wet season. During the dry season, shallow, isolated water bodies develop aquatic communities that are characterized by high turbidity because of the density of bacteria and algae as well as black coloration in the water from humic materials released during decomposition. Nevertheless, these temporary aquatic ecosystems have no endemic or even rare species. This is because of the likelihood of extinction during occasional very dry periods followed by the certainty of the extensive mixing among aquatic and wetland ecosystems that follows flooding in the wet season.

The seasonal wet-dry cycle provides a wide range of ecosystem services of great ecological and economic value. There were a large number of indigenous people who cultivated wild water rice, hunted deer, and constructed artificial islands within the swamps on which to live (Moss 1998). Water supply, food production, and waste processing all still contribute important economic values to the region, and benthic communities play an important role. The Pantanal is characterized by a density of large vertebrates and unique food webs (Heckman 1998). In addition to providing habitat to endangered terrestrial and semi-aquatic species such as the spotted jaguar (*Panthera onca*), giant anteater (*Myrmecophaga tridactyla*), and giant river otter (*Pteronura braziliensis*), this vast region supports more than 40 species of wading birds and more than 400 species of fish. These species have high "existence values" for many people, and they maintain the complex pelagic and benthic food webs that are the basis of many ecosystem services. Recreational uses are also of major economic importance. Hunters, fishermen, and conservationists travel from all over the world to view and to exploit this exceptional biodiversity.

During the dry season, much of the unflooded matrix within the Pantanal wetland becomes a savanna used for grazing large herds of cattle that are supported by nutrient cycling in the sediments. The savanna is divided into a series of cattle ranches (based on several million zebu cattle and local breeds) that have been burned regularly by ranchers for the last 150 years. Deforestation by burning to create more grazing land has led to soil erosion and high rates of sedimentation (see Covich et al., Chapter 3 and Ineson et al., Chapter 9). In more recent years, nonnative grasses have been introduced to improve forage, and pesticides and fertilizer use has increased in an effort to support the growth of rice and soybeans. These increased chemical inputs have negative effects—such as bioaccumulation and eutrophication—as occur elsewhere in the world (Moss 1998). Gold mining predates ranching by about 100 years, and open-pit gold mines are still being established. Purification of gold ore utilizes mercury, which is then evaporated, and there is some evidence of mercury pollution affecting birds (as in the Everglades). Subregions within the basin (such as Nhecolandia, Brazil, the second largest of 11 subregions) are being studied with remote sensing to increase available data on land-use values. Meanwhile, threats from the watershed have also increased, and the rivers feeding the Pantanal now introduce chemical contaminants, nutrients, and sediment from increasing urban developments, agricultural operations, and mines. Seidl and Moraes (2000) estimate the annual total value of ecosystem goods and services to the Nhecolandia subregion is more than US$15.5 billion.

The conflict is clear in the Pantanal between the maintenance of natural provisioning, supporting, and enriching services and the increase of the delivery of artificial services through agriculture and exploitation of natural resources. It is quite remarkable that such a diverse and unusual animal assemblage exists, given that it is dependent on a food chain with a very important benthic base that itself is by no means unique in its diversity or nature. How long the Pantanal ecosystems can continue to provide the wealth of ecosystem services under the increasing effects of the various stressors is unknown.

Nutrient Cycling and Productivity of Lakes: Lake Mendota, Wisconsin, United States

Lakes are used for a variety of ecosystem services, but because of their enclosed nature and the slow turnover of their water, they are often susceptible to a variety of threats, among them the loss of ecosystem services and resulting disservices. Eutrophication, for instance, results in rapid growth of blue-green algae, which affect tastes and odors of drinking water. Algal blooms disrupt filtration processes during water abstraction and can be toxic, which may affect drinking water for municipalities. These issues stimulated intensive experimental and modeling studies to determine whether such changes were irreversible (Baerenklau et al. 1999; Wilson & Carpenter 1999).

Lake Mendota in Madison, Wisconsin, is one of the most thoroughly studied medium-sized (approximately 4,000 ha) lakes in North America (Kitchell 1992). In the early 1980s, the combined decline of walleye populations and lost recreational fishing, together with concerns over unpredictable eruptions of noxious and sometimes toxic blue-green algae (cyanobacteria), led to a research effort demonstrating that water quality and food web management could be integrated. The management processes developed here and elsewhere were based on the trophic cascade concept (Carpenter & Kitchell 1993), in which enhanced populations of top piscivorous predators that feed on planktivorous fish led indirectly to the reduction of algal densities through the release of zooplankton populations that feed on algae. This approach of using one ecosystem function (predation) to enhance another (herbivory) and hence increase an ecosystem service (provision of high-quality water) has been utilized in a number of countries. In the case of Lake Mendota, management issues to solve conflicting service provision included: (1) trade-offs between increased stocking for walleye and northern pike fishing or managing for bass or perch (distinct "goods" for different people); (2) effects of increased water clarity (following removal of algae by grazing zooplankton) on deep light penetration, which can result in increased growth of submerged aquatic plants (that provide critical habitat for juvenile fishes, but can become weedy and reduce dissolved oxygen in the littoral zone during late summer and winter when the dead plants decay); and (3) disadvantages of improved water quality (clear water with lower concentrations of dissolved nutrients), which made it difficult to fulfill the demand for recreational fishing.

These integrative studies led to new questions about how management can enhance ecosystem services in freshwater bodies: How are "distinct" ecosystems, with apparently clearly defined surface boundaries (e.g., small ponds, large lakes, and rivers), interconnected hydrologically over time and space? How might these linked ecosystems function and affect each other in predictable ways? Why must fisheries biologists add fertilizers to increase fish production in some locations (hatcheries, aquaculture ponds) when water-quality engineers are designing treatment plants to remove nutrients in other "downstream" locations (groundwaters, rivers, and lakes)? Is production of fish for food versus recreation a necessary trade-off? Or can aquatic ecosystems be managed to optimize complex production functions? Can natural processes of nutrient cycling and organic-matter breakdown provide supplemental services that could save construction of new treatment plants? Answers to such questions have emphasized that sedimentary deposits and the species that live in these substrates are key regulators of nutrient cycling and productivity of different forms of plants. These basic elements of nutrient cycling (bottom-up control) interact with the effects of open-water predators such as fishes (top-down control) to jointly influence entire food webs (Kitchell 1992; Carpenter & Kitchell 1993).

The Everglades:
Coping with Heavy Metals and Ecosystem Disservice

The Everglades is a vast freshwater wetland that originally covered an area of more than 10,000 km^2 in south Florida, United States. It is part of a 100-km-long basin in which water flows along a gradual gradient of 3 cm/km from shallow Lake Okechobee to the mangroves lining Florida Bay. Exploitation of rich organic soils for agriculture, drainage for urban development, the construction of canals, and the impoundment of surface water for flood control and water storage have led to dramatic changes in flooding and fire regimes and nutrient inputs to the wetland. Because of draining and modifications in hydrologic regime, the area of the Everglades is now, in 2004, 35 percent of its original size.

The Everglades provides numerous ecosystem services for human well-being. Even in its much altered state, the Everglades filter polluted runoff from agricultural fields, yielding fresh, clean water for a variety of uses, including support of the estuarine ecosystems at its terminus. It harbors and produces a great quantity and diversity of wildlife, most notably alligators, crocodiles, the Florida panther, manatees, and a rich variety of aquatic birds. The fresh water it supplies to Florida Bay comes in a quality, quantity, and pattern of delivery that enables coastal ecosystems to provide their own suite of services. Finally, the Everglades provide aesthetic values, including recreation, to an audience that extends well beyond the boundaries of the United States.

Among the many changes to the Everglades that alerted scientists and resource managers to potential "ecosystem disservices," one that was particularly difficult to diagnose was the increase in concentrations of mercury in several species of vertebrates. Mercury contamination has been particularly pronounced for Everglades sport fishes; high levels have been detected in other vertebrates, including alligators, wading birds, and the Florida panther (Fink et al. 1999). Fear arose that agricultural pollutants introduced primarily into the north and eastern ends of the wetland were finding their way into and up the food chain, instigating a closer look at the mercury cycle in the Everglades. In fact, the emerging pattern of cause-and-effect is complex and, in some ways, very difficult to counteract.

Most of the mercury introduced to the Everglades comes from atmospheric sources, not from agriculture (Fink et al. 1999). Although some is from natural sources, such as volcanoes and outgassing from oceans, approximately 95 percent of the atmospheric mercury is released with coal combustion, waste incineration, and industrial processing (Krabbenhoft et al. 2003). Mercury in the atmosphere is primarily elemental mercury, which is relatively inert. Once deposited, it is subject to conversion to the more toxic methylmercury, a process performed primarily in an anoxic environment by sulfur-reducing bacteria, which are responsible for much of the organic carbon decomposition in the Everglades' sediments. An unusual feature in the Everglades' food chains is the dominance of periphyton over phytoplankton as the base of food chains (Browder et al. 1994). Periphyton is an assemblage of algae, bacteria, and associated microfauna that form a mat that overlies the surface sediments and often includes filamentous blue-green algae. Both mercury and methylmercury accumulate in periphyton, but it is still unclear how mercury becomes so concentrated in fishes near the top of the food chain. Complex interactions that change seasonally with fish diets (which include benthic invertebrates) and are affected by wetting cycles, fire, sunlight, total mercury concentrations, sulfate concentration, and levels of anoxia remain to be clarified (Gilmour et al. 1998;

Krabbenhoft et al. 2003). It is clear that in areas of nutrient enrichment, accumulation of biomass (often attributed to increased abundance and rates of growth of *Typha lati-filia* and *T. domingensis*) increases, rates of microbial activity and decomposition increase, and there is an increased tendency for mercury methylation (Gilmour et al. 1998).

Through our use of the atmosphere to perform the service of waste mercury disposal, humans are compromising animal and human food chains in the Everglades. Atmospheric deposition of mercury to the Everglades is approximately double the rate in rural Wisconsin for example, but it is difficult to determine the source of this input. Although it may be possible to manipulate Everglades water levels and mercury release patterns to minimize formation of methylmercury, the parts of this wetland that are most affected are the parts with the most natural fire and water regimes. Maintaining an environment that can continue to produce sustainable populations of sport fish and wildlife may not be compatible with atmospheric release of waste mercury. This example of the Everglades' ecosystem demonstrates the extent to which freshwater systems are often compromised by the use of ecosystem services in other realms.

Clean Drinking Water: Managing the Catskill Mountains of New York City's Watershed to Provide High-Quality Water Supplies

One of the major success stories in the use of natural ecosystems to deliver vital ecosystem services is the use of a series of river-reservoir ecosystems located in the Catskill Mountains to provide water for New York City's nearly nine million people (Ashendorff et al. 1997). Three large reservoir systems (Croton, Catskill, and Delaware) containing 19 reservoirs, 3 controlled lakes, and numerous tributaries cover an area of 5,000 km^2 with a reservoir capacity of 2.2×10^9 m^3. The US Environmental Protection Agency issued a "filtration avoidance status" in 1997 for five years in response to the city's request to upgrade their watershed management and enhance the capacity of natural ecosystems to maintain clean water. To avoid the potential expense of US\$2–8 billion over 10 years to build new, larger filtration plants to meet drinking water standards, the city invested US\$1–1.5 billion to restore natural ecosystem processes in the watershed (Ashendorff et al. 1997; Foran et al. 2000). The city agreed to construct a filtration plant if natural processes failed to meet EPA standards. Filtration is viewed as essential because chlorination is not completely effective in killing pathogens, particularly when there are high levels of suspended materials (Schoenen 2002).

New York City increased the capacity for natural nutrient retention and lower erosion by protecting riparian buffer zones along rivers and around reservoirs. Road construction within 30 m of a perennial stream and 15 m for an intermittent stream was prohibited. Non-point sources of nutrients and pesticides from stormwater runoff, septic tanks, and agricultural sources were also regulated. Water managers continued to monitor for protozoans, such as *Crytosporidium parvum* and *Giardia lamblia,* that cause cryptosporidiosis and giardiasis.

The city is now expected to save some US$300 million annually that would be necessary to run new filtration plants. The investment in natural capital reduced risks of contaminants, and the city can now focus on minimizing disinfectants at the final treatment stages. Although chlorination of drinking water is widely used, it can produce carcinogenic byproducts (e.g., chloroform, trihalomethane, and 260 other known chemicals) in drinking water, especially in ecosystems with high levels of organic matter (Zhang & Minear 2002).

Increasing effectiveness of natural ecosystem processes by watershed protection, restoration, and riparian management provides an example of how planners can cope with highly variable inputs that characterize this catchment (e.g., Frei et al. 2002). Boston, Seattle, San Francisco, and Greenville, South Carolina, are other examples where natural ecosystem services are used in conjunction with water treatment plants to ensure high-quality drinking water (O'Melia et al. 2000). This final case study illustrates the potential value of maintaining and enhancing natural ecosystem functioning in order to provide our vital ecosystem services.

Discussion and Conclusions

These case studies provide a spectrum of examples that demonstrate not only the ways in which ecosystem services are provided by freshwater benthic species, but also how they are vulnerable to human activities. The studies also provide some lessons that can be carried over into creating improved management of complex, interconnected ecosystems. A central feature of vulnerability of these benthic species is that, although freshwater is widely available, it is often extremely and unevenly distributed. Consequently, there are significant geographical disparities in the frequency and intensity of threats to the benthic biota and their associated ecosystem services. The trade-offs between ecological and economic values that are facing managers will be drastically different in the arid zones of Africa or India than in the arid western United States.

Trade-offs can be complex in wet or dry regions when exploitation of an ecosystem for one service eventually becomes a disservice relative to other needs. When managing for optimizing one service entails obstructing or even destroying the capacity to enjoy another service, either from the same ecosystem or from another ecosystem, planners must rationalize benefits from each service as well as the possibility of mitigation of effects in advance.

Managers often focus on a single problem and then seek to enhance a single ecosystem service to resolve the problem. For instance, designating the Catskill Mountains as a protected watershed for supplying New York City with fresh water provides a complex case study for other cities to consider. Will this approach establish a sustainable system for obtaining potable water without other unintended consequences? It is not clear what the effects of deflecting inflows for New York City's use will have on the Hudson River. Will the complexity of habitats that would have supported a greater number and diversity of fish and benthic infauna be affected by this alteration of flow? Will saline waters move farther upstream on the tidally altered portions of the Hudson River during droughts and thus affect other water supplies or certain benthic species' roles in providing other needed ecosystem services? Another poignant example of our inability to manage "single-service contracts" with freshwater aquatic ecosystems is the increased mercury contamination in the Everglades, now the scene of dramatic and expensive efforts to restore the suite of ecosystem services that it once provided. Restoring and preserving watersheds, redirecting wastewater to specially constructed wetland ecosystems, and guarding against the introduction of alien species are important goals—but complete analysis also requires comprehensive studies of inputs from the airshed. Mercury contamination from rainfall containing metals derived from burning fossil fuels persists as a major issue even if water pollution and hydrology can be managed to sustain the benthic biota. The Pantanal provides a positive example of a vast and complex landscape that continues to sustain high productivity in a mixture of wetland ecosystems that change shape and chemistry as wet and dry seasons alternate. Although the Pantanal is probably a fragile collection of interdependent ecosystems, and important parts may yet be lost to the threats that impinge on it, its example impresses on us the reality that assaults to a benthic community may ultimately be repairable. This is the hope for many of the severely stressed freshwater systems in Asia and Africa that have lost most of their most important natural provisioning and support ecosystem services (especially provision of potable clean water) through excessive inputs of pollutants. Generally, freshwater ecosystems are resilient to many kinds of short-term threats, once the perturbation stops and recovery becomes possible (Resh et al. 1988; Jansson et al. 1999). Much of this resilience and resistance can be attributable to the benthic community, which seems to provide a stabilizing interface between the physical environment and the nonbenthic community, and hence many of the services the freshwater ecosystems provide.

Risk analyses, to help balance our demands on valuable ecosystems more effectively, depend on the knowledge of what human activities are damaging, how such damage can be avoided, and the extent to which ecosystem services that are currently impaired can be restored. In order to offer such advice, we also need information on what governs the production of ecosystem services, the role of biodiversity in the sustainability and level of the services, and how production of services changes under altered conditions. Can we rank the threats in order to provide some guidance as to what actions are the most important to avoid and what are the most useful for beneficial restoration? We might speculate that geomorphic alteration is the most serious, as freshwater systems are not resilient to this sort of change. Chemical pollution and local extinctions can be more easily mitigated against and recovery is usually rapid. Large-scale watershed/catchment perturbations (such as changing land use) and resultant hydrochemical changes are far more significant again than local point source pollution. Invasive species may or may not be significant, depending on how they interact with the native communities and the scale of activity and population growth. Pulse disturbances (which occur over a reasonably limited spatial and temporal scale), be they hydrological (such as natural drought or flood) or chemical (such as pollution events), have limited long-term impact due to the high resilience of most freshwater systems. On the other hand, longer-term directed press disturbances (such as acidification, eutrophication, human-induced climate change, and hydrologic regulation) will have a greater impact on the ecological communities and hence on the provision of ecosystem services. The extent to which biodiversity provides some "insurance" against such changes is not clear at present, but the evidence from evolution suggests that some species may adapt to change while others may become extinct, thus, the provision of some ecosystem services may remain. For freshwater systems, however, this insurance is also at risk, as available information suggests that freshwater biodiversity has declined much faster over the past 30 years than either terrestrial or marine biodiversity. The greatest effects appear to be in the densely populated regions of the tropics (particularly South and Southeast Asia) and in dryland areas (Jenkins 2003). This complex linkage within and among ecosystems, like the example of the River Rhine, whose water quality has been improved enough to see the reappearance of many species of aquatic insects and fishes, will benefit from continued long-term monitoring and analysis of complex trade-offs inherent in management decisions.

Literature Cited

Admiraal, W., G. van der Velde, H. Smit, and W. Cazemier. 1993. The Rivers Rhine and Meuse in the Netherlands: Present state and signs of ecological recovery. *Hydrobiologia* 265:97–128.

Ashendorff, A., M.A. Principe, A. Seely, J. LaDuca, L. Beckhardt, W. Faber, Jr., and J. Mantus. 1997. Watershed protection for New York City's supply. *Journal of the American Water Works Association* 89:75–88.

Baerenklau, K.A., B. Stumborg, and R.C. Bishop. 1999. Non-point source pollution and present values: A contingent valuation study of Lake Mendota. *American Journal of Agricultural Economics* 81:1313.

Bronmark, C., and L-A. Hansson. 1998. *The Biology of Lakes and Ponds*. Oxford, Oxford University Press.

Browder, J.A., P.J. Gleason, and D.R. Swift. 1994. Periphyton in the Everglades: Spatial variation, environmental correlates and ecological implications. In: *Everglades: The Ecosystem and Its Restoration*, edited by S.M. Davis and J.C. Ogden, pp. 379–418. Delray Beach, Florida, St. Lucie Press.

Carpenter, S.R., and J.F. Kitchell. 1993. *The Trophic Cascade in Lakes*. Cambridge, UK, Cambridge University Press.

Cohen, J. 1995. *How Many People Can the Earth Support?* New York, W.W. Norton.

Covich, A.P., M.C. Austen, F. Bärlocher, E. Chauvet, B.J. Cardinale, C.L. Biles, P. Inchausti, O. Dangles, M. Solan, M.O. Gessner, B. Statzner, B. Moss. 2004. The role of biodiversity in the functioning of freshwater and marine benthic ecosystems: Current evidence and future research needs. *BioScience* 54:767–775.

Covich, A.P., M.A. Palmer, and T.A. Crowl. 1999. The role of benthic invertebrate species in freshwater ecosystems. *BioScience* 49:119–127.

Dynesius, M., and C. Nilsson. 1994. Fragmentation and flow regulation of river systems in the northern 3rd of the world. *Science* 266: 753–762.

Engelman, R., and P. LeRoy. 1993. *Sustaining Water: Population and the Future of Renewable Water Supplies.* Washington, DC, Population Action International.

Engleman, R., and P. LeRoy. 1995. *Sustaining Water: An Update.* Washington, DC, Population Action International.

Fink, L., D. Rumbold, and P. Rawlik. 1999. The Everglades mercury problem. In: *Everglades Interim Report,* pp. 7-1–7-71. West Palm Beach, Florida, South Florida Water Management District.

Fitzsimons, J.D., J.H. Leach, S.J. Nepzy, and V. Cairns. 1995. Impacts of zebra mussel on walleye reproduction in western Lake Erie. *Canadian Journal of Fisheries and Aquatic Sciences* 52:578–586.

Foran, J., T. Brosnan, M. Connor, J. Delfino, J. DePinto, K. Dickson, H. Humphrey, V. Novotny, R. Smith, M. Sobsey, and S. Stehman. 2000. A framework for comprehensive, integrated, waters monitoring in New York City. *Environmental Monitoring and Assessment* 62:147–167.

Frei, A., R.L. Armstrong, M.P. Clark, and M.C. Serreze. 2002. Catskill Mountain Water resources: Vulnerability, hydroclimatology, and climate-change sensitivity. *Annals of the Association of American Geographers* 92:203–224.

Giller, P.S., H. Hillebrand, U.-G. Berninger, M. Gessner, S. Hawkins, P. Inchausti, C. Inglis, H. Leslie, B. Malmqvist, M. Monaghan, P. Morin, and G. O'Mullan. 2004. Biodiversity effects on ecosystem functioning: Emerging issues and their experimental test in aquatic communities. *Oikos* 104:423–436.

Giller, P.S., and B. Malmqvist. 1998. *The Biology of Streams and Rivers.* Oxford, Oxford University Press.

Gilmour, C.C., G.S. Riedel, M.C. Ederington, J.T. Bell, J.M. Benoit, G.A. Gill, and M.C. Stordal. 1998. Methylmercury concentrations and production rates across a trophic gradient in the northern Everglades. *Biogeochemistry* 40:327–345.

Gleick, P.H. 2001. Safeguarding our water: Making every drop count. *Scientific American* 284:40–45.

Gregory, S.V., F.J. Swanson, W.A. McKee, and K.W. Cummins. 1991. An ecosystem perspective of riparian zones. *BioScience* 41:540–551.

Heckman, C.W. 1998. *The Pantanal of Poconé.* Boston, Kluwer Academic Publications.

Jansson, A., C. Folke, and J. Rockstrom. 1999. Linking freshwater flows and ecosystem services appropriated by people: The case of the Baltic Sea drainage basin. *Ecosystems* 2:351–366.

Jansson, R., C. Nilsson, M. Dynesius, and E. Andersson. 2000. Effects of river regulation on river-margin vegetation: A comparison of eight boreal rivers. *Ecological Applications* 10:203–224.

Jenkins, C.N., and S.L. Pimm. 2003. How big is the global weed patch? *Annals of the Missouri Botanical Garden* 90:172–178.

Jenkins, M. 2003. Prospects for biodiversity. *Science* 302:1175–1177.

Jungwirth, M., S. Muhar, and S. Schmutz. 2002. Re-establishing and assessing ecological integrity in riverine landscapes. *Freshwater Biology* 47:867–887.

Karatayev, A., L. Burlakova, and D. Padilla. 1997. The effects of *Deissena polymorpha* (Pallas) invasion on aquatic communities in Eastern Europe. *Journal of Shellfish Research* 16:187–203.

Keddy, P.A. 2000. *Wetland Ecology: Principles and Conservation.* Cambridge, UK, Cambridge University Press.

Kelleher, B., P.J.M. Berges, F.W.B. van den Brink, P.S. Giller, G. van de Velde, and A.B. de Vaate. 1998. Effects of exotic amphipod invasions on fish diet in the Lower Rhine. *Archiv fur Hydrobiologie* 143:363–383.

Kitchell, J.F., editor. 1992. *Food Web Management. A Case Study of Lake Mendota.* New York, Springer-Verlag.

Kitchell, J.F., D.E. Schindler, R. OgutuOhwayo, and P.N. Reinthal. 1997. The Nile perch in Lake Victoria: Interaction between predation and fisheries. *Ecological Applications* 7:653–664.

Krabbenhoft, D.P., W.H. Orem, G. Aiken, and C. Kendall. 2003. Aquatic cycling of mercury in the Everglades. http://sofia.usgs.gov/projects/evergl_merc/.

Lake, P.S., M.A. Palmer, P. Biro, J. Cole, A.P. Covich, C. Dahm, J. Gilbert, W. Goedkoop, K. Martens, and J. Verhoeven. 2000. Global change and the biodiversity of freshwater ecosystems: Impacts on linkages between above-sediment and sediment biota. *BioScience* 50:1099–1107.

Lelek, A., and C. Kohler. 1990. Restoration of fish communities of the river Rhine two years after a heavy pollution wave. *Regulated Rivers* 5:57–60.

Ludyanskiy, M.L., D. McDonald, and D. McNeill. 1993. Impact of the zebra mussel, a bivalve invader. *BioScience* 43:533–544.

MacIsaac, H.J. 1996. Potential abiotic and biotic impacts of zebra mussels on the inland waters of North America. *American Zoologist* 36:287–299.

Malmqvist, B., and G. Englund. 1996. Effects of hydropower-induced flow perturbations on mayfly (*Ephemeroptera*) richness and abundance in north Swedish river rapids. *Hydrobiologia* 341:145–158.

Malmqvist, B., and S. Rundle. 2002. Threats to the running water ecosystems of the world. *Environmental Conservation* 29:134–153.

Mason, C. 1996. *Biology of Freshwater Pollution,* 2nd edition. London, Longman Group.

Micklin, P.P. 1992. The Aral crisis: Introduction to the special issue. *Post Soviet Geography* 33:269–282.

Minchin, D., C. Maguire, and R. Rosell. 2003. The zebra mussel (*Deissena polymorpha* Palla) invades Ireland: Human mediated vectors and the potential for rapid international dispersal. *Biology and Environment: Proceedings of the Royal Irish Academy* 103B:23–30.

Moss, B. 1998. *The Ecology of Freshwaters. Man and Medium, Past to Future,* 3rd edition. Oxford, Blackwell Science.

Neumann, D. 2002. Ecological rehabilitation of a degraded large river system: Considerations based on case studies of macrozoobenthos and fish in the Lower Rhine and its catchment area. *International Review of Hydrobiology* 87:139–150.

Newman, S., J. Schuette, J.B. Grace, K. Rutchey, T. Fontaine, K.R. Reddy, and M. Pietrucha. 1998. Factors influencing cattail abundance in the northern Everglades. *Aquatic Botany* 60:265–280.

O'Melia, C.R., M.J. Pfeffer, P.K. Barten, G.E. Dickey, M.W. Garcia, C.N. Haas, R.G. Hunter, R.R. Lowrance, C.L. Moe, C.L. Paulson, R.H. Platt, J.L. Schnoor, T.R. Schueler, J.M. Symons, and R.G. Wetzel. 2000. *Watershed Management for Potable Water Supply. Assessing the New York City Strategy.* Washington, DC, National Research Council, National Academy Press.

Paine, R.T., and T.M. Zaret. 1973. Species introduction in a tropical lake. *Science* 182:449–455.

Palmer, M.A., A.P. Covich, B.J. Finlay, J. Gibert, K.D. Hyde, R.K. Johnson, T. Kairesalo, S. Lake, C.R. Lovell, R.J. Naiman, C. Ricci, F. Sabater, and D. Strayer. 1997. Biodiversity and ecosystem processes in freshwater sediments. *Ambio* 26:571–577.

Poff, L., and J.V. Ward. 1989. Implications of streamflow variability and predictability for lotic community structure: A regional analysis of streamflow patterns. *Canadian Journal of Fisheries and Aquatic Sciences* 46:1805–1818.

Por, F.D. 1995. *The Panatal of Mato Grosso (Brazil): World's Largest Wetlands. Monagraphie Biologicae. V. 73.* Dordrecht, The Netherlands, Kluwer Academic.

Postel, S., and S. Carpenter. 1997. Freshwater ecosystem services. In: *Nature's Services: Societal Dependence on Natural Ecosystems,* edited by G.C. Daily, pp. 195–214. Washington, DC, Island Press.

Power, M.E., W.E. Dietrich, and J.C. Finlay. 1996. Dams and downstream aquatic biodiversity: Potential food web consequences of hydrologic and geomorphic change. *Environmental Management* 20:887–895.

Power, M.E., A. Sun, G. Parker, W.E. Dietrich, and J.T. Wootton. 1995. Hydraulic food chain models. *BioScience* 45:159–167.

Resh, V.H., A.V. Brown, A.P. Covich, M.E. Gurtz, H.W. Li, and G.W. Minshall. 1988. The role of disturbance in stream ecology. *Journal of the North American Benthological Society* 7:433–455.

Rosenberg, D.M., P. McCully, and C.M. Pringle. 2000. Global-scale environmental effects of hydrological alterations: Introduction. *BioScience* 50:746–751.

Schoenen, D. 2002. Role of disinfection in suppressing the spread of pathogens with drinking water: Possibilities and limitations. *Water Research* 3:3874–3888.

Seidl, A.F., and A.S. Moraes. 2000. Global valuation of ecosystem services: Application to the Pantanal da Nhecolandia, Brazil. *Ecological Economics* 33:1–6.

Simons, J.H.E.J., C. Bakker, M.H. Schropp, L.H. Jans, F.R. Kok, and R.E. Grift. 2001. Man-made secondary channels along the River Rhine (the Netherlands): Results of post-project monitoring. *Regulated Rivers—Research and Management* 17:473–491.

Singer, P.C. 2002. *Relative Dominance of Haloacetic Acids and Trihalomethanes in Treated Drinking Water.* Denver, Colorado, American Water Works Association.

Sioli, H. 1984. The Amazon and its main affluents: Hydrography, morphology of the river courses and river types. In: *The Amazon. Limnology and Landscape Ecology of a Mighty Tropical River and Its Basin,* edited by H. Sioli, pp. 127–166. Dordrecht, The Netherlands, Dr W. Junk Publishers.

Stoddard, J.L., D.S. Jeffries, A. Lukewille, T.A. Clair, P.J. Dillon, C.T. Driscoll, M. Forsius, M. Johannessen, J.S. Kahl, J.H. Kellogg, A. Kemp, J. Mannio, D.T. Monteith, P.S. Murdoch, S. Patrick, A. Rebsdorf, B.L. Skjelkvale, M.P. Stainton, T. Traaen, H. van Dam, K.E. Webster, J. Wieting, and A. Wilander. 1999. Regional trends in aquatic recovery from acidification in North America and Europe. *Nature* 401:1389–1395.

Strayer, D.L. 1999. Effects of alien species on freshwater mollusks in North America. *Journal of the North American Benthological Society* 18:74–98.

van den Brink, F.W.B. 1994. *Impact of Hydrology on Floodplain Lake Ecosystems Along the Lower Rhine and Meuse.* Den Haag, The Netherlands, CIP-Gegevens Koninklijke Bibliotheek.

van den Brink, F.W.B., G. van der Velde, and W. Cazemeir. 1990. The faunistic composition of the freshwater section of the river Rhine in the Netherlands: Present state and changes since 1900. In: *Biologie des Rheins*, edited by R. Kinzelbach and G. Friedrich, pp. 1:192–216. Stuttgart, Germany, Limnologie aktuell.

van den Brink, F.W.B., G. van der Velde, and A. Bij de vaste. 1993a. Ecological aspects, explosive range extension and impact of a mass invader, *Corophium curvispinum* Sars, 1895 (Crustacea: Amphipoda), in the Lower Rhine (The Netherlands). *Oecologia* 93:224–232.

van den Brink, F.W.B., B. Paffen, F. Oosterbroek, and G. van der Velde. 1993b. Immigration of *Echinogammarus* (Stebbing 1899) (Crustacea: Amphipoda) into the Netherlands via the Lower Rhine. *Bulletin Zoologisch Museum, University Van Amsterdam* 13(14):167–169.

van der Velde, G., F.W.B. van den Brink, R. Van der Gaag, and P. Bergers. 1990. Changes in numbers of mobile macroinvertebrates and fish in the river Waal in 1987, studied by sampling the cooling water intakes of a power plant: First results of a Rhine biomonitoring project. In: *Biologides Rheins*, edited by R. Kinzelbach and G. Friedrich, pp. 1:326–342. Stuttgart, Germany, Limnologie aktuell.

van der Velde, G., B. Paffen, and F.W.B. van den Brink. 1994. Decline of zebra mussel populations in the Rhine: Competition between two mass invaders (*Dreissena polymorpha* and *Corophium curvispinum*). *Naturwissenschaften* 81:32–34.

Van de Weijden, C., and J. Middleburg. 1989. Hydrogeochemistry of the River Rhine: Long term and seasonal variability, elemental budgets, base levels and pollution. *Water Research* 23:1247–1266.

Van Urk, G., and H. Smit. 1989. The Lower Rhine: Geomorphological changes. In: *Historical Changes in Large Alluvial Rivers: Western Europe*, edited by G. Petts, pp. 167–182. New York, John Wiley and Sons.

Vörösmarty, C., P. Greenm, J. Salisbury, and R. Lammers. 2000. Global water resources: Vulnerability from climate change and population growth. *Science* 289:284–288.

Wall, D.H., P.V.R. Snelgrove, and A.P. Covich. 2001. Conservation priorities for soil and sediment invertebrates. In: *Conservation Biology*, edited by M.E. Soulé and G.H. Orians. Washington, DC, Island Press.

Ward, J.V., and D. Garcia de Jalon. 1991. Ephemeroptera of regulated mountain streams in Spain and Colorado. In: *Overview and Strategies of Ephemeroptera and Plecoptera*, edited by J. Alba-Tercedor and A. Sanchez-Ortega, pp. 576–578. Gainesville, Florida, Sandhill Crane Press.

Ward, J.V., K. Tochner, and F. Scheimer. 1999. Biodiversity of riparian floodplain river ecosystems: Ecotones and connectivity. *Regulated Rivers, Research and Management* 15:125–139.

Wilson, M.A., and S.R. Carpenter. 1999. Economic valuation of freshwater ecosystems services in the United States: 1971–1997. *Ecological Applications* 9:772–783.

World Commission on Environment and Development. 1987. *Our Common Future*. Oxford, Oxford University Press.

Zhang, X., and R.A. Minear. 2002. Characterization of high molecular weight disinfection by products resulting from chlorination of aquatic humic substances. *Environmental Science and Technology* 36:4033–4038.

7

Vulnerability of Marine Sedimentary Ecosystem Services to Human Activities

Paul V.R. Snelgrove, Melanie C. Austen, Stephen J. Hawkins, Thomas M. Iliffe, Ronald T. Kneib, Lisa A. Levin, Jan Marcin Weslawski, Robert B. Whitlatch, and James R. Garey

Marine sedimentary ecosystems encompass more of the Earth's surface than any other habitat, but many people consider the sea floor to be a vast, monotonous environment that is remote from human disturbance. Biodiversity is often thought to be of little consequence to the resources we extract from the ocean, to the health of the marine environment, and to quality of human life. Nonetheless, marine sediments provide important extractable goods such as fisheries. They also play regulatory roles in global transfer and cycling of materials and energy (see Weslawski et al., Chapter 4). Many of the ecosystem processes (*sensu* Chapin et al. 2002) that occur in marine sediments also have important consequences for the sustainability of ecosystem services valued by human society (e.g., shoreline stabilization, waste recycling, etc.). Marine sediments from the highly visible coastline (Figure 7.1) to the remote and lesser known deep sea vary in exposure to threats and the probability of being harmed by threats if exposed; both issues contribute to vulnerability, defined here as the propensity of ecological systems to suffer harm from exposure to external stresses and shocks (Wall et al. 2001; Folke et al. 2002). The goal of this chapter is to examine potential threats to biodiversity in marine sedimentary ecosystems. We ask whether changes in biodiversity will increase the vulnerability of systems to loss of processes and services, and we examine their potential to recover from impacts. In most examples, loss of services provided by marine sediments has not yet been addressed by scientists; the discussion, therefore, focuses on processes with the underlying implication that these affect services (see Weslawski et al., Chapter

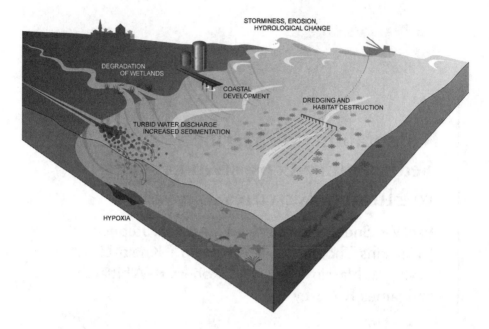

Figure 7.1. Schematic summary of the major threats to ecosystem goods and services in estuarine and coastal ecosystems.

Table 7.1. Summary of major threats to marine sedimentary systems and the scales at which they are manifested.

Scale	Issue	Ecosystems Most Affected
Small Scale (individual embayments)	invasive species	estuaries, wetlands
	disease	estuaries, wetlands
	coastal development/ habitat alteration	wetlands, estuaries
Mid Scale (regional scale)	hydrologic alteration	wetlands, estuaries, intertidal
	overfishing/habitat destruction	shelf, slope, estuaries, wetlands
	eutrophication/pollution	estuaries, shelf
Large Scale (basin scale)	climate change, including:	
	sea-level change	wetlands, intertidal, estuaries
	rainfall patterns	wetlands, intertidal
	temperature	shelf, intertidal
	wind & circulation	all
	salinity	estuaries, wetlands, intertidal
	ultraviolet radiation	all

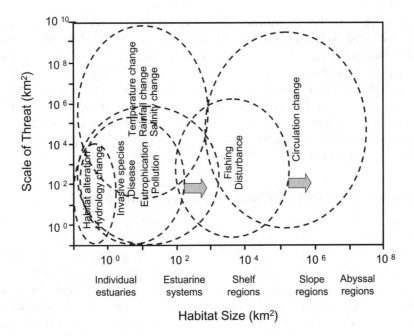

Figure 7.2. Scales of threats to marine sedimentary habitats. Circles denote ranges of affected area and habitats. Arrows for invasive species/disease and fishing disturbance indicate potentially larger scales of threat with increased human disturbance.

4). As in Chapter 4 of this volume, marine systems here are grouped into estuarine, continental shelf, and deep-sea sediments. *Estuaries* encompass sedimentary habitat at the land-sea interface where seawater is measurably diluted by freshwater input, the *continental shelf* refers to the sea floor between continents and the top of the continental slope (~130 m deep), and *deep-sea habitat* encompasses the comparatively steep (~4°) continental slope that extends from the edge of the continental shelf to the continental rise (~ 4,000 m) that grades into the vast abyssal plains (4,000–6,000 m) that cover much of the deep ocean floor. *Seamounts* are submerged mountains on abyssal plains that extend thousands of meters above the sea floor.

Threats and Scales of Vulnerability

Sedimentary ecosystems are exposed to multiple threats stemming from human activity (Table 7.1) with the potential to exert significant impacts on species composition and ecosystem processes across local (a bay or semi-enclosed coastal area), regional (hundreds of kilometers) and basin-wide scales (Hixon et al. 2001). Like other ecosystems, marine environments have a capacity to withstand and recover from human-induced disturbances. For example, sustainable fisheries are possible because populations of organisms have an innate capacity to increase their numbers, and individuals removed by fishing are replaced by the offspring of those that escaped the fishery. Likewise, many marine systems can accommodate some sewage input without loss of biodiversity or generation of anoxia. The problem is that disturbances often exceed the capacity of the system to recover, resulting in loss of species and, in some instances, loss of the capacity to produce goods and services. The size of a system and its proximity to human populations may influence vulnerability at different scales, and we therefore examine vulnerability to threats by scale of impact (Figure 7.2) from the land-sea interface to the deep sea.

Local-Scale Threats (kilometers to tens of kilometers)

Estuaries are more likely than continental shelf (gently sloping regions from 0–130 m depth between the shoreline and the upper edge of the continental slope) and deep sea (regions beyond the continental shelf, including the continental slope [~130–3,000 m], continental rise [~3,000–4,000 m], and the abyssal plains [~4,000–6,000 m]) systems to experience severe local-scale effects because they are smaller in spatial extent, less open to adjacent systems, and physically closer to human populations (Levin et al. 2001). However, we also recognize that intense local effects (e.g., many individual trawls) can spread and become regional (e.g., broad-scale fishing impacts). Below we outline local-scale threats.

Alien Taxa

Alien species are often introduced from point sources (e.g., ship ballast water and hull fouling, mariculture [marine aquaculture] activities; see Carlton & Geller 1993), though in instances where species are highly invasive, they can quickly create problems at a regional scale through rapid dispersal of propagules (Wehrmann et al. 2000). Invasive species may also exhibit population lags over many generations before becoming a problem (Crooks & Soulé 1999). Within estuarine habitats, biotic invasions rank second only to habitat alteration among potential threats to biodiversity (Vitousek et al. 1997; Carlton 2001), and invasion rates to coastal marine systems have accelerated over the past 200 years (Ruiz et al. 2000; Carlton 2001). Most benthic faunal invaders have been crustaceans and mollusks (Ruiz et al. 2000), but polychaetes (Roehner et al. 1996), plants, and disease organisms also have altered benthic diversity (Carlton 2001). Introduced plants can alter communities and ecosystem processes because of their trophic importance in estuarine food webs and effects on physical habitat structure (e.g., architecture, sedimentation). For example, *Phragmites australis* reduces plant species diversity where it invades (Lenssen et al. 2000), lowering soil salinity and water levels, reducing microtopographic features, and changing sediment oxidation (Windham & Lathrop 1999), but it is not known to reduce benthic faunal diversity in freshwater wetlands (Ailstock et al. 2001). Invasions can reduce density, elevate species richness, and change infaunal composition in tidal wetlands (Talley & Levin 2001) with unknown functional consequences. Hybridization with local species can also alter or reduce genetic diversity (Ayres et al. 1999).

Disease

Diseases represent a natural threat to all living organisms. Humans can exacerbate that threat by increasing susceptibility to disease through physiological stress, by accelerating the spread of disease, or by introducing contaminants into the marine environment in an attempt to control disease. Effects may include loss of goods and services or alteration of processes. For example, chemical contamination can cause chronic lesions in (Moore et al. 1997) and increased effects of parasitism (Khan 1987) on benthic fishes, reducing the commercial value and potentially compromising the sustainability of fisheries. Transport of toxic dinoflagellate cysts in ballast water can contaminate new areas (Hallegraeff & Bolch 1992), resulting in losses to mariculture and wild shellfish yields. Antibiotic use in mariculture appears to be a localized threat, but drugs used to control parasites may have toxic effects on benthic invertebrates that extend well beyond mariculture locales (Goldburg et al. 2001). Many mariculture programs also use non-endemic species or stocks, and introductions or escapes of pathogens have contributed to invasive species problems in estuaries (Carlton 2001). However, there is little evidence linking this emerging industry to large-scale disease impacts on benthic communities (Rothschild et al. 1994; Naylor et al. 2000).

Coastal Development and Habitat Alteration

Human activities alter the physical structure of coastal habitats across a range of spatial and temporal scales. Lerberg et al. (2000) related the amount and type of shoreline development to changes in species richness and populations of key functional groups. Perhaps the most direct negative impacts on infaunal (organisms living in sediments) communities are associated with sediment disturbance during dredging of waterways for navigation (Newell et al. 1998), and the more chronic effects of heavy gear (e.g., fish trawls, scallop dredges) frequently used to harvest estuarine and coastal species (Hall-Spencer & Moore 2000; Thrush & Dayton 2002). Even relatively localized sediment disturbances (e.g., pipeline installation, increased land erosion runoff of sediments) can affect macrofaunal biodiversity (Lewis et al. 2002). These disturbances often damage or eliminate larger, sessile, and long-lived benthic filter-feeding invertebrates (e.g., corals, sponges, bivalves), as well as seagrass, mangrove, and marsh vegetation. This reduces structural complexity and habitat diversity as well as the system's capacity to trap sediments (Morris et al. 2002), improve water quality (Coen et al. 1999), and mitigate effects of stressors such as hypoxia (depleted oxygen concentrations) (Lenihan & Peterson 1998). In finfish pen mariculture, hypoxia beneath pens reduces benthic diversity (Weston 1990). Disturbances in natural habitats including estuaries often create conditions that favor different species than those that occur in undisturbed areas, with potential consequences for ecosystem processes and their associated benefits to humans.

Channel alteration—construction of impoundments, docks, roads, and shoreline armoring (to protect nearshore property from erosion)—may affect the functioning of coastal habitats (Kneib 2000). Shoreline armoring, where protective physical structures such as concrete breakwaters are built to protect coastal property, prevents natural inland migration of wetlands with sea-level rise (Pethick 2001), posing a serious long-term threat to both diversity and ecosystem processes (Morris et al. 2002). Armored shorelines support hard substrate faunas (i.e., organisms that are found on exposed bedrock and other nonsedimentary habitat) in environments that previously supported sedimentary ecosystems (Davis et al. 2002). Key trophic links and ecosystem processes (e.g., energy flows) also depend on maintaining physical corridors between estuaries and habitats for movement of materials and organisms (Micheli & Peterson 1999; Kneib 2000).

Estuaries function as nurseries for mobile species that are harvested elsewhere (e.g., nearshore shelf waters) and are sometimes spared the direct impacts of commercial (though not recreational) fisheries. Yet there are many examples of overexploitation of resources in estuaries and bays including reductions or extinctions of benthic filter-feeders, a key functional group that strongly influences bentho-pelagic coupling (linkages between the bottom [benthic] and water column [pelagic] environment) and water quality (Jackson et al. 2001; Dayton et al. 2002).

Regional-Scale Threats (hundreds to thousands of kilometers)

Hydrological Alteration

Hydrological alteration occurs when rivers are diverted or outflow is substantially reduced for other purposes (e.g., hydroelectric projects, irrigation). Effects often are local (such as single estuaries) but significant diversions can affect an entire coastal region. Increased extraction and consumption of water lowers the water table and reduces freshwater input, increasing marine influences on estuaries with potentially serious consequences for benthic biodiversity and ecosystem processes. Changes in the amount or timing of freshwater flow disrupt the balance between the influence of land and sea on the estuarine environments (Sklar & Browder 1998), and affect the life cycles of species dependent upon seasonal cycles of freshwater inputs to estuaries. Regional rainfall correlates with the areal extent of intertidal marshes and mangrove forests within temperate and tropical estuaries and coastal embayments (Deegan et al. 1986), and species diversity in estuaries is related to changes in salinity (Kalke & Montagna 1991; Jassby et al. 1995). Pulsed flooding events can change sediment and nutrient inputs to favor deposit-feeders (high sediment loads) and suspension-feeders (low sediment loads) in different seasons (Salen-Picard & Arlhac 2002). Likewise, reduced ocean exchange can create hyper- or hyposalinity and hypoxia (Teske & Wooldridge 2001).

Mariculture

Studies of broad-scale impacts of aquaculture on biodiversity of benthic communities are rare, but environmental impacts of mariculture operation are linked to all of the threats to biodiversity listed above, but especially to habitat destruction. In many developing countries, intertidal mangrove forest and other wetlands have been replaced by shrimp mariculture ponds. In Thailand, 54 percent of the estuarine mangrove forests present in 1961 were converted to other uses by 1993, primarily for the construction of shrimp ponds (Macintosh et al. 2002). The high biodiversity of benthic communities and high production of wild shrimp and other fisheries' species in natural mangrove forest systems may be permanently lost (Naylor et al. 2000) because attempts to restore these damaged habitats rarely produce the biodiversity of crustacean and molluscan species found in undisturbed mature mangrove habitats (Macintosh et al. 2002).

Overfishing and Habitat Alteration

Shallow coastal and shelf systems are often fished heavily for target species with considerable collateral damage through bycatch. Excessive fishing mortality of target species and/or bycatch can alter food webs substantially (Pauly et al. 1998). Bottom trawl fisheries can also be destructive by homogenizing large areas of sea floor that provide habitats for benthic and near-bottom species (Dayton et al. 1995). Although overfishing is concentrated in estuarine and shelf areas (Auster et al. 1996), it has spread to seamounts and continental slopes.

Fishing impacts are the threat of greatest concern on shelf systems (National Research Council 1995). In estuarine coastal and shelf systems, overfishing reduces stock levels and alters trophic support processes and food web dynamics between the pelagic and benthic zones and also within the benthos (Pauly & MacLean 2003). Commercially harvested fishes are often top predators, and their overexploitation has cascading effects through to lower parts of the food chain (Myers & Worm 2003; Worm & Myers 2004). Similarly, harvesting of large filter-feeder bivalves eliminates populations that perform key processes in benthic systems (Pauly et al. 1998; Jackson et al. 2001). Food web disruption through overfishing, eutrophication, and invasive species act in concert to further reduce availability of fish and shellfish as food (Lancelot et al. 2002).

Although recreational fisheries can cause substantial mortality in top-level predators (Dayton et al. 2002), the effect on benthic biodiversity is less studied and findings are mixed. There is evidence from manipulative mesocosm experiments (Kneib

1991; Duffy & Hay 2000) that indicates trophic interactions involving top predators can have cascading effects on the composition (hence biodiversity) of benthic assemblages, but it is uncertain whether mesocosm results can be scaled up to any large, open system. It is nearly impossible to investigate this issue in many regions because of public and political resistance to establishing fully protected marine reserves that prohibit fisheries exploitation. For example, less than 1 percent of marine environments in the United States are more than nominally protected; most of these are coral habitats (Palumbi 2002).

At larger scales offshore and inshore, habitat alteration is caused by dragging fishing gear through sediments, which disrupts established chemical and biotic gradients. Globally, trawling impacts many thousands of square kilometers of continental shelf seabed, although effects depend on sediment type, fishing gear, and trawling frequency (Collie et al. 2000). The larger macrofauna and epifauna living within and on the sediment are most vulnerable to trawling mortality (Kaiser et al. 2000), injury from fishing gear (Ramsay et al. 1998), and exposure to predators. Mollusks such as whelks are physically rolled over and exposed by fishing gears and are, therefore, more vulnerable to predation (Ramsay & Kaiser 1998). Scavengers increase food intake in fished areas (Kaiser & Spencer 1994), and sharks and rays associate trawlers with food (Stevens et al. 2000). Larger infauna and epifauna create local, small-scale habitat patchiness through altering water flow or creating biogenic structures such as burrows and tubes, or by moving sediment and feeding. By reducing these populations, trawling homogenizes the sediment and landscape interconnection via habitat patches and refugia (Auster et al. 1995; Thrush et al. 2002). Organisms that create small-scale heterogeneity and habitat structure also stabilize sediment through accretion and alteration of water flow at the sediment-water interface.

The great depth of the abyssal plains, their remoteness, and the low abundance of harvestable species concentrates most deep-sea fisheries activities in upper to mid continental slope depths. Deep-sea organisms often grow slowly, mature at a later age than shallow water species, and produce comparatively few offspring (Merritt & Haedrich 1997). Declines in catch rate and stock size have been observed for commercially exploited fish species (Clark 1995), deep-sea crab, and shrimp (Orensanz et al. 1998). Bycatch has also caused drastic reductions in abundances of deep-sea fishes that are not targeted by fisherman in the North Atlantic (Baker & Haedrich 2003). As in other systems, fishing removes top predators that are hypothesized to play a regulatory cropping role in deep-sea diversity maintenance (Dayton & Hessler 1972; Myers & Worm 2003) and patch creation (Grassle & Sanders 1973). Fishing gear can have significant impacts on deep-sea bottom habitat akin to that seen in shelf environments (Koslow et al. 2000). Seamounts and deep-sea coral (*Lophelia*) reefs are an extreme example of this problem, where destructive trawl gear damages epifauna that are unique to specific seamount or reef regions and that may provide key habitat for fishes and other organisms (Koslow et al. 2001; Hall-Spencer et al. 2002).

In more homogeneous deep-sea settings, the impacts of habitat destruction and predator removal are mitigated by the vast extent of the environment. Effects of fisheries on deep-sea diversity (aside from seamounts with high levels of endemism) are difficult to know, given that estimates of the total species present varies by an order of magnitude (Grassle & Maciolek 1992; Lambshead & Boucher 2003) and distribution maps do not exist for most deep-sea species.

Removal of pelagic top consumers (e.g., whales, tuna) and the fishing down of food webs may also have cascading effects on deep-water food chains (Butman et al. 1995). Inputs of large organic matter falling to the deep-sea floor may decrease, affecting food supply for many scavenger species that contribute to diets of other deep-water fishes.

The deep sea holds extensive mineral and hydrocarbon deposits. Manganese nodules (rock-like, golf ball–sized structures that are rich in manganese and other minerals and are found in dense concentrations in some areas of the abyssal plains) and crusts, polymetalic sulfides (chimney-like deposits that form at deep-sea hydrothermal vents), and phosphorites all contain valuable cobalt, nickel, copper, and manganese but their extraction is not yet economically viable (Glover & Smith 2003). Potential effects on benthos depend in part on whether mining waste (mostly sediments) is discharged at the seabed. Any scenario includes damage and crushing from mining gear, but seabed discharge could also smother organisms. Some effects included initial reduction followed by increases in megafaunal abundance driven primarily by scavenging species (Bluhm et al. 1995) and no change or decreases in macrofaunal abundance and diversity (Borowski 2001). Although benthic populations can recover from simulated disturbance within three years, diversity effects may remain after seven years. Experiments show relatively localized, noncatastrophic effects, though the small scale of these experiments suggests caution in extrapolating to commercial impacts (Ozturgut et al. 1981). There is, nonetheless, a specialized manganese nodule fauna that could be reduced or lost in intense mining scenarios (Thiel et al. 1993).

Oil and gas drilling beyond the shelf break has gone from only a possibility, a half century ago, to a reality in recent decades. For example, the Brazilian company PetroBos has some production activity at depths greater than 1,800 m off western Africa (Glover & Smith 2003), and exploration by other companies is occurring at depths greater than 3,000 m in the Gulf of Mexico (Minerals Management Service 2004). If drilling is done carefully, effects on diversity and biomass may be very localized (kilometer scale), with the exception of greater effects in connection with major spills (Thiel 2003) or drilling over long temporal scales. In contrast, the mining of methane hydrates, though not yet technically possible, could destabilize slope areas and cause mass slumping events with broad-scale mortality (Thiel 2003). Methane hydrate is a gas that freezes at depths greater than 300 m, and may someday be the most important deep-sea resource because it is thought that oceans hold over twice as much carbon in methane hydrate as all other sources of fossil fuel (USGS Survey Fact sheet: http://marine.usgs.gov/fact-sheets/gas-hydrates/title.html).

Eutrophication and Pollutants

Estuaries and semi-enclosed bays often support dense human populations and industry, with associated high inputs of pollutants (e.g., heavy metals, hydrocarbons) and nutrients, which can cause phytoplankton blooms that subsequently decay and create bottom hypoxia. Estuaries and bays with limited exchange with the open ocean are particularly vulnerable; however, larger-scale hypoxic events are becoming increasingly frequent over shelf areas adjacent to inputs from nutrient-enriched estuaries or large river systems (Diaz & Rosenberg 1995). Although many estuarine and shelf ecosystems have a significant capacity to recycle organic matter and nutrients without inducing hypoxia, a balanced community of microbes and bioturbators is needed to provide this ecosystem service.

Pollutant inputs to estuaries and coasts from point and non-point sources are an increasing global problem (Boesch et al. 2001). Many estuaries are exposed to a broad spectrum of contaminants, including oil spills that induce complex changes in benthic invertebrate assemblages (Long 2000; Peterson 2000; Peterson et al. 2003). Excessive nutrient inputs to estuaries have cascading effects that reduce benthic biodiversity (Howarth et al. 2000), and increase frequency of harmful algal blooms and hypoxia or anoxia (Boesch et al. 2001). Reductions in rooted vegetation (Howarth et al. 2000) and simplification of community structure result from increased frequency and persistence of hypoxic events (Diaz & Rosenberg 1995).

Effluents from mariculture may contribute to eutrophication and local changes in sedimentation, all of which affect benthic biodiversity and the functional role of benthos in semi-enclosed ecosystems such as estuaries, bays, and fjords (Naylor et al. 2000). Few and limited effects of mariculture effluents have been measured in the water column (McKinnon et al. 2002), and most impacts have been seen in sediments (Ervik et al. 1997). Sedimentation of excess food particles and fecal material in the vicinity of mariculture facilities has contributed to local hypoxia/anoxia and reductions in benthic biodiversity. As long as an impacted area is not already over-enriched from other sources, the additional nutrient loading is not expected to have substantial large-scale impacts on benthic biodiversity. Rearing filter-feeding organisms in mariculture may even improve water quality and offset negative effects of eutrophication on benthic biodiversity. Modestly sized mariculture operations designed to minimize habitat destruction and, with a focus on production of native species low in the food web (bivalves and herbivorous fishes), may pose little threat to benthic biodiversity (Naylor et al. 2000).

On the continental shelves and through the interface between estuarine and shelf waters, threats include direct and indirect inputs of pollutants (Clark 1997). Some pollutants are discharged directly onto the shelf via pipelines or dumping from ships. Indirect inputs occur through discharge into rivers and estuaries, and through airborne contamination of rainwater. Hypoxia can stress and increase mortality in meio- and macrofauna (Gray et al. 2002), and in larger organisms such as fishes (Diaz & Rosenberg 1995). The resulting disruption of microbial communities and production affects oxygenation processes and detoxification of contaminants within the sediments. Dumping of waste in deeper waters of the shelf and spillage from oil installations and ships also result in pollution farther from the coastal margins. Shipwrecks and collisions cause large-scale pollution (Peterson 2000; Peterson et al. 2003) with effects that can persist for many years (e.g., more than 10 years in Dauvin 1998), though the spatial scales of spills in the open ocean tend to be small relative to the habitat area, and cumulative effects are unknown.

Dumping dredged material causes habitat alteration through smothering and toxicity (Somerfield et al. 1995). Alteration of water flow into rivers and dredging of river channels changes shelf hydrology, salinity, and sediment deposition (de Jonge & de Jonge 2002). Effects may be even more severe than in estuaries because shelf organisms seem less physiologically tolerant to these types of disturbance.

Materials dumped in the deep sea have included conventional munitions and chemical weapons (Schriever et al. 1997), low- and intermediate-level radioactive wastes (Thiel 2003), sewage sludge (Bothner et al. 1994), dredge spoil containing contaminants (Tyler 2003), and various vessels and structures associated with the military, shipping, and oil and gas exploitation (Schriever et al. 1997). Modern deep-sea research has revealed that this ecosystem is more dynamic and reactive than previously believed (Tyler 2003) and for the most part, western countries have ceased deep-ocean dumping. The deep sea is now being considered as a repository for excess carbon dioxide (Herzog et al. 2000), which is liquid at high pressures and low temperatures (Glover & Smith 2003). Deep-sea carbon dioxide disposal could mitigate atmospheric CO_2 increases and decrease surface ocean pH, but concurrently reduce deep-ocean pH (Caldeira & Duffy 2000) to lethal levels in the CO_2 plumes (Glover & Smith 2003).

Continental slopes exhibit the highest carbon deposition, the greatest animal productivity, and perhaps the highest biodiversity in the deep sea (Rex 1983). Because they are also the most accessible deep-sea setting, human activities can have greater impacts on biodiversity and key slope sediment processes such as carbon burial or production. There are few studies that address these issues for the deep sea.

Enrichment experiments in the deep sea (Snelgrove et al. 1992) suggest that organic input selects for shorter-lived, surface-dwelling, opportunistic species, and similar effects may occur with disposal of sewage sludge and dredged materials in the deep sea. Enrichment can also alter diets and locally enhance epibenthic species (Grassle 1991; Van Dover et al. 1992). Physical disturbance from bottom trawling also generates opportunistic, low-diversity assemblages. Replacement of larger, deeper-dwelling species by small opportunists reduces bioturbation, limits oxygenation of the sediment, and slows carbon degradation and burial rates. Slope macrofauna typically consume and bury (to 10 cm or more) fresh organic matter within days of its arrival on the seabed (Graf 1989; Levin et al. 1997), so that most labile carbon is respired or buried rapidly. At abyssal depths, human-induced suppression of bioturbation and carbon burial may occur from CO_2 sequestration or nodule mining activities, with effects akin to natural mass slumping (Masson et al. 1994).

Broad-Scale Threats (Basin Scale): Climate Change

Climate change will alter salinity, temperature, and wind patterns, which will affect local, regional, and broad-scale hydrography (Manabe et al. 1994), but this may occur slowly. Sea-level rise, for example, will have few direct effects on shelf or deep-sea systems, but it will have substantial effects on coastal environments (Smith et al. 2000). Similarly, rising sea levels cause a loss of open tidal flats and landward migration of marshes (Donnelly & Bertness 2001), but shoreline armoring constrains this movement, resulting in intertidal estuarine habitat loss (Pethick 2001). Because rooted vegetation is a key part of habitat structure and primary production in estuarine wetlands, broad-scale losses associated with sea-level rise (Barras et al. 2003) pose serious threats to biodiversity and ecosystem processes and services (Morris et al. 2002).

At larger spatial scales, impacts will probably result from changes in the frequency of basin-wide meteorological phenomena such as ENSO (El Niño-Southern Ocean) and NAO (North Atlantic Oscillation) events and in the strength of boundary currents.

Climate changes that alter seasonal patterns of rainfall or temperature have little effect on deep-sea systems. However, in coastal regions, climate change may cause local extinctions of endemic benthic invertebrates adapted to historical patterns of environmental variation, and at the same time promote expansion of the range of some invasive species. Although many invasive species in estuaries are euryhaline (able to tolerate a wide range of salinities), more invasions occur in high than low salinities (Ruiz et al. 2000). Reduced freshwater flow allows marine waters to penetrate farther into estuaries, and successful introductions, including those of disease organisms (e.g., the parasitic dinoflagellate, *Hematodinium* spp.), are more frequent during droughts or reduced freshwater flows into the system (Carlton et al. 1990; Messick et al. 1999). There is apparently wide variation in the resistance of estuarine benthic communities to invasion (Ruiz et al. 2000), and evidence from hard substrate communities suggests that diversity helps resist invasion (Stachowicz et al. 2002). With climatic change, it is therefore reasonable to expect shifts in invasion resistance and the abundance of invaders.

Climate change is increasing the frequency and scale of extreme weather, and strong winds increase wave action, which results in physical disturbance to the sea bed along coastlines and in shallower shelf waters (Hall 1994). The effects are similar to those of widespread fishing disturbance. In shallow regions, wind-induced wave action may actually increase nutrient cycling due to physical disturbance of the sediment, enhancing pollutant detoxification through increased oxygenation of sediments. Under normal conditions, shelf benthic organisms in the Peru-Chile upwelling zone are often oxygen stressed, with low biodiversity, low biomass, and little bioturbation activity. El Niño upwelling replaces normally hypoxic shelf waters with well-oxygenated water masses, promoting greater benthic biodiversity, productivity, nutrient cycling, and food production. This type of climate event may enhance most marine goods and services (Tarazona et al. 1988; Gutierrez et al. 2000).

Additional rainfall during winter and reduced rainfall during summer in temperate zones will change coastal hydrologic characteristics and could affect sediment loading and the effectiveness of flood and erosion control by reef-forming and sediment-stabilizing organisms. Accelerated glacial meltdown, wide-scale reductions in salinity (some of which will be offset by increased damming of rivers), and increased temperatures will change benthic community structure and diversity (Smith et al. 2000; Austen et al. 2002). Increases in temperature may enhance productivity where nutrients are not limiting, resulting in positive effects on some of the goods and services provided by marine benthos.

Little is known about potential effects of increased ultraviolet radiation in marine sedimentary benthos, but work on pelagic eggs of near-bottom fishes (Kouwenberg et al. 1999) suggest that effects on planktonic larval stages are possible.

Comparison Among Systems

We have summarized the vulnerability to different threats for each of the three ecosystem groupings (Tables 7.A1–7.A3 in the appendix on page 184), as they relate to provisioning of goods (e.g., food, fiber) and services (e.g., water filtration, flood control, waste recycling), as well as their supporting ecosystem processes (e.g., carbon sequestration, nutrient cycling, decomposition), habitat maintenance services, and aesthetic services (spiritual enrichment, recreation, scientific inquiry). Because many of the available data on vulnerability are either anecdotal or collected and interpreted in very different frameworks for different threats, an objective, quantitative comparison across systems is not possible. We have used our collective experiences and available studies on ecosystem processes and services to develop a qualitative ranking scheme to address the relative vulnerability of different systems to loss of services or processes (Tables 7.A1–7.A3 in the appendix on page 184). In some instances a given threat may actually enhance some service or process, and thus a negative score is possible. For example, an introduced species may create ecological havoc but, if it is edible and abundant, it may increase provisioning of food.

We interpret from this exercise that estuarine systems are currently the most vulnerable of our three broadly categorized marine systems, in part because of the wide range of services that are carried out in the environments and in part because of the intensity and number of threats. Remote deep-sea systems are the least vulnerable marine systems under current patterns of exploitation because of their large area of interconnected habitat and their relatively low exposure to human activity. Because of its size, large- to mid-scale effects (e.g., climate change) are of greatest concern. Nonetheless, seamounts represent an area of extreme concern, and continental slopes are vulnerable to human exploitation where localized effects are increasing.

Recovery, Restoration, and Rehabilitation of Marine Ecosystems

Marine restoration has lagged behind that of terrestrial and freshwater ecosystems, in part because oceans are massive in scale and common in ownership, which hinders intervention. In recent years, however, a framework for recovery, restoration, and rehabilitation (see Frid & Clark 1999 and Hawkins et al. 1999 for definitions) of marine ecosystems has developed, particularly on coastal systems dominated by habitat-providing biota such as seagrass beds, mangroves, and salt marshes (Ewel et al. 2001).

In open marine ecosystems, barriers to larval dispersal are few and there is good potential for natural recovery by recolonization from unimpacted populations. Whereas natural recovery can be rapid, active restoration in open ocean systems is difficult except where biological structure creates and maintains habitat. For example, biologically generated structure such as seagrasses, saltmarsh halophytes, and mangroves may not disperse well (Orth et al. 1994), but active planting can accelerate habitat restoration and associated ecosystem processes and services.

Enclosed waters such as estuaries, bays, and lagoons can be amenable to restoration by the manipulation of water quality and ocean flushing. For example, macrophytes can sequester nutrients in semi-enclosed areas that might otherwise become eutrophic. Phytoplankton standing crop may also be influenced by filter feeders (Officer et al. 1982; Davies et al. 1989; Hily 1991). The openness of most marine systems means that water quality can be improved by regional reductions of harmful inputs and activities. Where flushing by ocean tides is restricted, for example, by road construction or episodic inlet closure through sedimentation, habitat may be restored by active dredging, manipulating flow, and constructing permanently open inlets.

Many nearshore and shelf areas are heavily impacted by disturbances where effects can be mitigated only by leaving large areas undisturbed. Marine protected areas are proposed worldwide as pragmatic precautionary fisheries management tools with broader marine conservation benefits (US Commission on Ocean Policy 2004). These can have various levels of protection (see Jennings & Kaiser 1998 and Hall 1999 for excellent overviews) from absolute exclusion ("no take zones") to less strict regions where gear types are limited or fishing is excluded in some seasons.

Most coastal restoration work has focused on particular habitats, biotopes, assemblages, or species, but coastal ecosystems are strongly interconnected and coordinated efforts are sometimes needed. Thus, seagrass restoration may aid saltmarsh recovery or restoration. Active restoration has the strongest cascading effects where "ecosystem engineers " (Lawton 1994) are involved. Oyster reefs, mussel beds, seagrass beds, saltmarsh, and mangroves are all examples where a strong structural element is conferred by the dominant biota. Rooted macrophytes (e.g., seagrasses) can be planted to form the nuclei of new beds (Fonseca et al. 2002). Conversely, oyster reef restoration is less successful near salt marsh and seagrass beds, which provide corridors for large mobile predators (Micheli & Peterson 1999).

Conclusion

Our discussion focused on many processes and services in marine sediments that in some instances are only now being recognized. The importance of critical habitats, such as those utilized by juvenile fishes, has become a major focus for fisheries organizations only in the last decade or so. Are there future services and processes that are deteriorating but are receiving no attention because they are not yet recognized? It is the unknown future value of marine sedimentary systems that places a particular urgency on preserving the remaining systems that are still relatively pristine; the potential losses in service and process that we have summarized here may represent only part of the story. Ultimately, biodiversity has value in and of itself that goes beyond goods and services provided to humans and the processes that biodiversity may help to support. But even for those who fail to recognize a beauty in the diversity of living things and our ethical obligation to preserve this diversity, the potential for loss of desirable goods and services supported by marine sedimentary fauna should at least provide pause (and provide food) for thought, and an impetus to protect these living resources.

Literature Cited

Ailstock, M.S., C.M. Norman, and P.J. Bushmann. 2001. Common reed *Phragmites australis:* Control and effects upon biodiversity in freshwater nontidal wetlands. *Restoration Ecology* 9:49–59.

Austen, M.C., P.J.D. Lambshead, P. Hutchings, G. Boucher, C. Heip, G. King, I. Koike, C.R. Smith, and P.V.R. Snelgrove. 2002. Biodiversity links above and below the marine sediment-water interface that may influence community stability. *Biodiversity and Conservation* 11:113–136.

Auster, P.J., R.J. Malatesta, R.W. Langton, L. Watling, P.C. Valentine, C.L.S. Donaldson, E.W. Langton, A.N. Shepard, and I.G. Babb. 1996. The impacts of mobile fishing gear on seafloor habitats in the Gulf of Maine (northwest Atlantic): Implications for conservation of fish populations. *Reviews in Fisheries Science* 4:185–202.

Auster, P.J., R.J. Malatesta, and S.C. LaRosa. 1995. Patterns of microhabitat utilization by mobile megafauna on the southern New England (USA) continental shelf and slope. *Marine Ecology Progress Series* 127:77–85.

Ayres, D.R., D. Garcia-Rossi, H.G. Davis, and D.R. Strong. 1999. Extent and degree of hybridization between exotic (*Spartina alterniflora*) and native (*S. foliosa*) cordgrass (Poaceae) in California, USA determined by random amplified polymorphic DNA (RAPDs). *Molecular Ecology* 8:1179–1186.

Baker, K., and R.L. Haedrich. 2003. Could some deep-sea fishes be species-at-risk? International Deep-Sea Biology Conference, Aug. 2003, Coos Bay, Oregon (Abstract).

Barras, J., S. Beville, D. Britsch, S. Hartley, S. Hawes, J. Johnson, P. Kemp, Q. Kinler, A. Martucci, J. Porthouse, D. Reed, K. Roy, S. Sapkota, and J. Suhayda. 2003. Historical and predicted coastal Louisiana land changes 1978–2050. USGS Open File Report 03-334, 36 pp.

Bluhm, H., G. Schriever, and H. Thiel. 1995. Megabenthic recolonization in an experimental disturbed abyssal manganese nodule area. *Marine Georesources and Geotechnology* 13:393–416.

Boesch, D.F., R.H. Burroughs, J.E. Baker, R.P. Mason, C.L. Rowe, and R.L. Siefert. 2001. *Marine Pollution in the United States*. Arlington, Virginia: Pew Oceans Commission.

Borowski, C. 2001. Physically disturbed, deep-sea macrofauna in the Peru basin, southeast Pacific revisited 7 years after the experimental impact. *Deep-Sea Research* II 48:3809–3839.

Bothner, M.H., H. Takada, I. Knight, R. Hill, B. Butman, J.W. Farrington, R.R. Colwell, and J.F. Grassle. 1994. Sewage contamination in sediments beneath a deep-ocean dump site off New York. *Marine Environmental Research* 38:43–59.

Butman, C.A., J.T. Carlton, and S.R. Palumbi. 1995. Whaling effects on deep-sea biodiversity. *Conservation Biology* 9:462–464.

Caldeira, K., and P.B. Duffy. 2000. The role of the Southern Ocean in uptake and storage of anthropogenic carbon dioxide. *Science* 287:620–622.

Carlton, J.T. 2001. *Introduced Species in U.S. Coastal Waters: Environmental Impacts and Management Priorities*. Arlington, Virginia: Pew Oceans Commission.

Carlton, J.T., and J.B. Geller. 1993. Ecological roulette: The global transport of non-indigenous marine organisms. *Science* 261:78–82.

Carlton, J.T., J.K. Thompson, L.E. Schemel, and F.H. Nichols. 1990. Remarkable invasion of San Francisco Bay (California, USA) by the Asian clam *Potamocorbula amurensis* I. Introduction and dispersal. *Marine Ecology Progress Series* 66:81–94.

Chapin, F.S. III, P.A. Matson, and H.A. Mooney. 2002. *Principles of Terrestrial Ecosystems Ecology*. New York, Springer-Verlag.

Clark, M.R. 1995. Experience with management of orange roughy (*Hoplostethus atlanticus*) in New Zealand waters, and the effects of commercial fishing on stocks over the period 1980–1993. In: *Deep-Water Fisheries of the North Atlantic Oceanic Slope*, edited by A.G. Hopper, pp. 251–266. Dordrecht, The Netherlands, Kluwer Academic Publishers.

Clark, R.B., editor. 1997. *Marine Pollution, 4th Edition*. Oxford, Clarendon Press.

Coen, L.D., R.E. Giotta, M.W. Luckenbach, and D.L. Breitburg. 1999. Oyster reef function, enhancement, and restoration: Habitat development and utilization by commercially- and ecologically-important species. *Journal of Shellfish Research* 18:712.

Collie, J.S., S.J. Hall, M.J. Kaiser, and I.R. Poiner. 2000. Shelf sea fishing disturbance of benthos: Trends and predictions. *Journal of Animal Ecology* 69:785–798.

Crooks, J.A., and M.E. Soulé. 1999. Lag times in population explosions of invasive species: Causes and implications. In: *Invasive Species and Biodiversity Management*, edited by O.T. Sandlund, P.J. Schei, and A. Vikan, pp. 103–126. Boston, Kluwer Academic Publisher.

Dauvin, J.C. 1998. The fine sand *Abra alba* community of the Bay of Morlaix twenty years after the Amoco Cadiz oil spill. *Marine Pollution Bulletin* 36:669–676.

Davies, B.R., V. Stuart, and M. de Villiers. 1989. The filtration activity of a serpulid polychaete population (*Ficopomatus enigmaticus* Fauvel) and its effects on water quality in a coastal marina. *Estuarine Coastal and Shelf Sciences* 29:613–620.

Davis, J.L., L.A. Levin, and S. Walther. 2002. Artificial armored shorelines: Introduction of open coast communities into a southern California bay. *Marine Biology* 140:1249–1262.

Dayton, P.K., and R.R. Hessler. 1972. Role of biological disturbance in maintaining diversity in the deep sea. *Deep-Sea Research* 19:199–208.

Dayton, P.K., S.F. Thrush, M.T. Agardy, and R.J. Hofman. 1995. Environmental effects of marine fishing. *Aquatic Conservation: Marine and Freshwater Ecosystems* 5:205–232.

Dayton, P.K., S. Thrush, and F.C. Coleman. 2002. *Ecological Effects of Fishing in Marine Ecosystems of the United States.* Arlington, Virginia, Pew Oceans Commission.

Deegan, L.A., J.W. Day, Jr., J.G. Gosselink, A. Yáñez-Arancibia, G. Soberón Chávez, and P. Sánchez-Gil. 1986. Relationships among physical characteristics, vegetation distribution and fisheries yield in Gulf of Mexico estuaries. In: *Estuarine Variability*, edited by D.A. Wolfe, pp. 83–100. Orlando, Florida, Academic Press, Inc.

de Jonge, V.N., and D.J. de Jonge. 2002. "Global change" impact of inter-annual variation in water discharge as a driving factor to dredging and spoil disposal in the river Rhine system and of turbidity in the Wadden Sea. *Estuarine Coastal and Shelf Science* 55:969–991.

Diaz, R.J., and R. Rosenberg. 1995. Marine benthic hypoxia: A review of its ecological effects and the behavioural responses of benthic macrofauna. *Oceanography and Marine Biology: An Annual Review* 33:245–303.

Donnelly, J.P., and M.D. Bertness. 2001. Rapid shoreward encroachment of salt marsh cordgrass in response to accelerated sea-level rise. *Proceedings of the National Academy of Sciences, USA* 98:14218–14223.

Duffy, J.E., and M.E. Hay. 2000. Strong impacts of grazing amphipods on the organization of a benthic community. *Ecological Monographs* 70:237–263.

Ervik, A., P.K. Hansen, J. Aure, A. Stigebrandt, P. Johannessen, and T. Jahnsen. 1997. Regulating the local environmental impact of intensive marine fish farming, I. The concept of the MOM system (Modelling—Ongrowing fish farms—Monitoring). *Aquaculture* 158:85–94.

Ewel, K.C., C. Cressa, R.T. Kneib, P.S. Lake, L.A. Levin, M.A. Palmer, P. Snelgrove, and D.H. Wall. 2001. Managing critical transition zones. *Ecosystems* 4:452–460.

Folke, C., S. Carpenter, T. Elmqvist, L. Gunderson, C.S. Holling, and B. Walker. 2002. Resilience and sustainable development: Building adaptive capacity in a world of transformations. *Ambio* 31:437–440.

Fonseca, M., P.E. Whitfield, N.M. Kelly, and S.S. Bell. 2002. Modeling seagrass landscape pattern and associated ecological attributes. *Ecological Applications* 12:218–237.

Frid, C.L.J., and S. Clark. 1999. Restoring aquatic ecosystems: An overview. *Aquatic Conservation: Marine and Freshwater Ecosystems* 9:1–4.

Glover, A.G., and C.R. Smith. 2003. The deep-sea floor ecosystem: Current status and prospects for anthropogenic change by the year 2025. *Environmental Conservation* 30:1–23.

Goldburg, R.J., M.S. Elliott, and R.I. Naylor. 2001. *Marine Aquaculture in the United States: Environmental Impacts and Policy Options.* Arlington, Virginia, Pew Oceans Commission.

Graf, G. 1989. Benthic-pelagic coupling in a deep-sea benthic community. *Nature* 341:437–439.

Grassle, J.F. 1991. Effects of sewage sludge on deep-sea communities (abstract). *EOS, Transactions of the American Geophysical Union* 72:84.

Grassle, J.F., and N.J. Maciolek. 1992. Deep-sea species richness: Regional and local diversity estimates from quantitative bottom samples. *American Naturalist* 139:313–341.

Grassle, J.F., and H.L. Sanders. 1973. Life histories and the role of disturbance. *Deep-Sea Research* 20:643–659.

Gray, J.S., R.S.S. Wu, and Y.Y. Or. 2002. Effects of hypoxia and organic enrichment on the coastal marine environment. *Marine Ecology Progress Series* 238:249–279.

Gutierrez, D., V.A. Gallardo, S. Mayor, C. Neira, C. Vasquez, J. Sellanes, M. Rivas, A. Soto, F. Carrasco, and M. Baltazar. 2000. Effects of dissolved oxygen and fresh organic matter on the bioturbation potential of macrofauna in sublittoral sediments off Central Chile during the 1997/1998 El Nino. *Marine Ecology Progress Series* 202:81–99.

Hall, S. 1994. Physical disturbance and marine benthic communities: Life in unconsolidated sediments. *Oceanography and Marine Biology: An Annual Review* 32:179–239.

Hall, S. 1999. *The Effects of Fishing on Marine Ecosystems and Communities.* Oxford, Blackwell Sciences.

Hallegraeff, G.M., and C.J. Bolch. 1992. Transport of diatom and dinoflagellate resting spores in ships' ballast water: Implications for plankton biogeography and aquaculture. *Journal of Plankton Research* 14:1067–1084.

Hall-Spencer, J., V. Allain, and J.H. Fossa. 2002. Trawling damage to Northeast Atlantic ancient coral reefs. *Proceedings of the Royal Society of London, Series B* 269:507–511.

Hall-Spencer, J.M., and P.G. Moore. 2000. Scallop dredging has profound, long-term impacts on maerl habitats. *ICES Journal of Marine Science* 57:1407–1415.

Hawkins, S.J., J.R. Allen, and S. Bray. 1999. Restoration of temperate marine and coastal ecosystems: Nudging nature. *Aquatic Conservation: Marine and Freshwater Ecosystems* 9:23–46.

Herzog, H., B. Eliasson, and O. Kaarstad. 2000. Capturing greenhouse gases. *Scientific American* 282:72–79.

Hily, C. 1991. Is the activity of benthic suspension feeders a factor controlling water quality in the Bay of Brest? *Marine Ecology Progress Series* 69:179–188.

Hixon, M.A., P.D. Boersma, M.L. Hunter, Jr., F. Micheli, E.A. Norse, H.P. Possingham, and P.V.R. Snelgrove. 2001. Oceans at risk: Research priorities in marine conservation biology. In: *Conservation Biology: Research Priorities for the Next Decade,* edited by M.E. Soulé, and G.H. Orians, pp. 125–154. Washington, DC, Island Press.

Howarth, R., D. Anderson, J. Cloern, C. Elfring, C. Hopkinson, B. LaPointe, T. Malone, N. Marcus, K. McGlathery, A. Sharpley, and D. Walker. 2000. Nutrient pollution of coastal rivers, bays and seas. *Issues in Ecology* 7:1–15.

Jackson, J.B.C., M.X. Kirby, W.H. Berger, K.A. Bjorndal, L.W. Botsford, B.J. Bourque, R.H. Bradbury, R. Cooke, J. Erlandson, J.A. Estes, T.P. Hughes, S. Kidwell, C.B. Lange, H.S. Lenihan, J.M. Pandolfi, C.H. Peterson, R.S. Steneck, M.J. Tegner, and R.R. Warner. 2001. Historical overfishing and the recent collapse of coastal ecosystems. *Science* 293:629–638.

Jassby, A.D., W.J. Kimmerer, S.G. Monismith, C. Armor, J.W. Cloern, T.M. Powell, J.R. Schubel, and T.J. Vendlinski. 1995. Isohaline position as a habitat indicator for estuarine populations. *Ecological Applications* 5:272–289.

Jennings, S., and M.J. Kaiser. 1998. The effects of fishing on marine ecosystems. *Advances in Marine Biology* 34:201–352.

Kaiser M.J., K. Ramsay, C.A. Richardson, F.E. Spence, and A.R. Brand. 2000. Chronic fishing disturbance has changed shelf sea benthic community structure. *Journal of Animal Ecology* 69:494–503.

Kaiser, M.J., and B.E. Spencer. 1994. Fish scavenging behaviour in recently trawled areas. *Marine Ecology Progress Series* 112:41–49.

Kalke, R.D., and P.A. Montagna. 1991. The effect of freshwater inflow on macrobenthos in the Lavaca River delta and upper Lavaca Bay, Texas. *Contributions in Marine Science* 32:49–71.

Khan, R.A. 1987. Effects of chronic exposure to petroleum hydrocarbons on two species of marine fish infected with a hemoprotozoan, *Trypanosoma murmanensis*. *Canadian Journal of Zoology* 65:2703–2709.

Kneib, R.T. 1991. Indirect effects in experimental studies of marine soft-sediment communities. *American Zoologist* 31:874–885.

Kneib, R.T. 2000. Salt marsh ecoscapes and production transfers by estuarine nekton in the southeastern United States. In: *Concepts and Controversies in Tidal Marsh Ecology*, edited by M.P. Weinstein and D.A. Kreeger, pp. 267–292. Dordrecht; Boston, Kluwer Academic Publishers.

Koslow, J.A., G.W. Boehlert, J.D. Gordon, R.L. Haedrich, P. Lorance, and N. Parin. 2000. Continental slope and deep-sea fisheries: Implications for a fragile ecosystem. *ICES Journal of Marine Science* 57:548–557.

Koslow, J.A., K. Gowlett-Holmes, J.K. Lowry, T. O'Hara, G.C.B. Poore, and A. Williams. 2001. Seamount benthic macrofauna off southern Tasmania: Community structure and impacts of trawling. *Marine Ecology Progress Series* 213:111–125.

Kouwenberg, J.H.M., H.I. Browman, J.J. Cullen, R.F. Davis, J.-F. St-Pierre, and J.A. Runge. 1999. Biological weighting of ultraviolet (280–400 nm) induced mortality in marine zooplankton and fish. I. Atlantic cod (*Gadus morhua*) eggs. *Marine Biology* 134:269–284.

Lambshead, P.J.D., and G. Boucher. 2003. Marine nematode deep-sea diversity—hyperdiverse or hype? *Journal of Biogeography* 30:475–485.

Lancelot, C., J. Staneva, D. Van Eeckhout, J. Beckers, and E. Stanev. 2002. Modelling the Danube-influenced north-western continental shelf of the Black Sea. II: Ecosystem response to changes in nutrient delivery by the Danube River after its damming in 1972. *Estuarine, Coastal and Shelf Science* 54:473–499.

Lawton, J.H. 1994. What do species do in ecosystems? *Oikos* 71:367–374.

Lenihan, H.S., and C.H. Peterson. 1998. How habitat degradation through fishery disturbance enhances impacts of hypoxia on oyster reefs. *Ecological Applications* 8:128–140.

Lennsen, J.P.M., F.B.J. Menting, W.H. van der Putten, and C.W.P.M. Blom. 2000. Variation in species composition and species richness within *Phragmites australis* dominated riparian zones. *Plant Ecology* 147:137–146.

Lerberg, S.B., A.F. Holland, and D.M. Sanger. 2000. Responses of tidal creek macrobenthic communities to the effects of watershed development. *Estuaries* 23:838–853.

Levin, L.A., N. Blair, D.J. DeMaster, G. Plaia, W. Fornes, C. Martin, and C. Thomas. 1997. Rapid subduction of organic matter by *Maldanid polychaetes* on the North Carolina slope. *Journal of Marine Research* 55:595–611.

Levin, L.A., D.F. Boesch, A. Covich, C. Dahm, C. Erséus, K.C. Ewel, R.T. Kneib, A. Moldenke, M.A. Palmer, P. Snelgrove, D. Strayer, and J.M. Weslawski. 2001. The function of marine critical transition zones and the importance of sediment biodiversity. *Ecosystems* 4:430–451.

Lewis, L.J., J. Davenport, and T.C. Kelly. 2002. A study of the impact of a pipeline construction on estuarine benthic invertebrate communities. *Estuarine, Coastal and Shelf Science* 55:213–221.

Long, E.R. 2000. Degraded sediment quality in U.S. estuaries: A review of magnitude and ecological implications. *Ecological Applications* 10:338–349.

Macintosh, D.J., E.C. Ashton, and S. Havanon. 2002. Mangrove rehabilitation and intertidal biodiversity: A study in the Ranong mangrove ecosystem, Thailand. *Estuarine, Coastal and Shelf Science* 55:331–345.

Manabe, S., R.J. Stouffer, and M.J. Spelman. 1994. Response of a coupled ocean-atmosphere model to increasing atmospheric carbon dioxide. *Ambio* 23:44–49.

Masson, D.G., J.A. Cartwright, L.M. Pinheiro, R.B. Whitmarsh, M.O. Beslier, and H. Roeser. 1994. Compressional deformation at the ocean-continent transition in the NE Atlantic. *Journal of the Geological Society of London* 151:607–613.

McKinnon, A.D., L.A. Trott, M. Cappo, D.K. Miller, S. Duggan, P. Speare, and A. Davidson. 2002. The trophic fate of shrimp farm effluent in mangrove creeks of North Queensland, Australia. *Estuarine, Coastal and Shelf Science* 55:655–671.

Merritt, N.R., and R.L. Haedrich. 1997. *Deep-Sea Demersal Fish and Fisheries.* London, Chapman and Hall.

Messick, G.A., S.J. Jordan, and W.F. van Heukelem. 1999. Salinity and temperature effects on *Hematodinium* sp. in the blue crab *Callinectes sapidus. Journal of Shellfish Research* 18:657–662.

Micheli, F., and C.H. Peterson. 1999. Estuarine vegetated habitats as corridors for predator movements. *Conservation Biology* 13:869–881.

Minerals Management Service. 2004. *Deepwater Gulf of Mexico 2004: America's Expanding Frontier.* US Department of the Interior, Minerals Management Service, Gulf of Mexico OCS Region. OCS Report MMS 2004-021.

Moore, M.J., R.M. Smolowitz, and J.J. Stegeman. 1997. Stages of hydropic vacuolation in the liver of winter flounder *Pleuronectes americanus* from a chemically contaminated site. *Diseases of Aquatic Organisms* 31:19–28.

Morris, J.T., P.V. Sundarweshwar, C.T. Nietch, B. Kjerfve, and D.R. Cahoon. 2002. Responses of coastal wetlands to rising sea level. *Ecology* 83:2869–2877.

Myers, R.A., and B. Worm. 2003. Rapid worldwide depletion of predatory fish communities. *Nature* 423:280–283.

National Research Council. 1995. *Understanding Marine Biodiversity.* Washington, DC, National Academy Press.

Naylor, R.L., R.J. Goldburg, J.H. Primavera, N. Kautsky, M.C.M. Beveridge, J. Clay, C. Folke, J. Lubchenco, H. Mooney, and M. Troell. 2000. Effect of aquaculture on world fish supplies. *Nature* 405:1017–1024.

Newell, R.C., L.J. Seiderer, and D.R. Hitchcock. 1998. The impact of dredging works in coastal waters: A review of the sensitivity to disturbance and subsequent recovery of biological resources on the sea bed. *Oceanography and Marine Biology: An Annual Review* 36:127–178.

Officer, C.B., T.J. Smayda, and R. Mann. 1982. Benthic filter feeding: A natural eutroph-ication control. *Marine Ecology Progress Series* 9:203–210.

Orensanz, J.M., J. Amstrong, D. Armstrong, and R. Hilborn. 1998. Crustacean resources are vulnerable to serial depletion—the multifaceted decline of crab and shrimp fisheries in the Greater Gulf of Alaska. *Reviews in Fish Biology and Fishes* 8:117–176.

Orth, R.J., M. Luckenbach, and K.A. Moore. 1994. Seed dispersal in a marine macrophyte: Implications for colonization and restoration. *Ecology* 75:1927–1939.

Ozturgut, E., J.W. Lavelle, and R.E. Burns. 1981. Impacts of manganese nodule mining on the environment: Results from pilot-scale mining tests in the north equatorial Pacific. In: *Marine Environmental Pollution 2: Dumping and Mining*, edited by R.W. Geyer, pp. 437–474. Elsevier Oceanography Series. Amsterdam, Elsevier.

Palumbi, S.R. 2002. *Marine Reserves: A Tool for Ecosystem Management and Conservation.* Arlington, Virginia, Pew Ocean Commission.

Pauly, D., V. Christensen, J. Dalsgaard, R. Froese, and F. Torres. 1998. Fishing down marine food webs. *Science* 279:860–863.

Pauly, D., and J. MacLean. 2003. *In a Perfect Ocean: The State of Fisheries and Ecosystems in the North Atlantic Ocean.* Washington, DC, Island Press.

Peterson, C.H. 2000. The Exxon Valdez oil spill in Alaska: Acute, indirect and chronic effects on the ecosystem. *Advances in Marine Biology* 39:3–84.

Peterson, C.H., S.D. Rice, J.W. Short, D. Esler, J.L. Bodkin, B.E. Ballachey, and D.B. Irons. 2003. Long-term ecosystem response to the *Exxon Valdez* oil spill. *Science* 302:2082–2086.

Pethick, J. 2001. Coastal management and sea-level rise. *Catena* 42:307–322.

Ramsay K., and M.J. Kaiser. 1998. Demersal fishing disturbance increases predation risk for whelks (*Buccinum undatum* L.). *Journal of Sea Research* 39:299–304.

Ramsay K., M.J. Kaiser, and R.N. Hughes. 1998. Responses of benthic scavengers to fish-ing disturbance by towed gears in different habitats. *Journal of Experimental Marine Biology and Ecology* 224:73–89.

Rex, M.A. 1983. Geographic patterns of species diversity in the deep-sea benthos. In: *The Sea. Deep-sea Biology*, edited by G.T. Rowe, pp. 453–472. New York, John Wiley and Sons.

Roehner, M., R. Bastrop, and K. Juerss. 1996. Colonization of Europe by two American genetic types or species of the genus *Marenzelleria* (Polychaeta: Spionidae): An electrophoretic analysis of allozymes. *Marine Biology* 127:277–287.

Rothschild, B.J., J.S. Ault, P. Goulletquer, and M. Héral. 1994. Decline of the Chesapeake Bay oyster population: A century of habitat destruction and overfishing. *Marine Ecology Progress Series* 111:29–39.

Ruiz, G.M., P.W. Fofonoff, J.T. Carlton, M.J. Wonham, and A.H. Hines. 2000. Invasion of coastal marine communities in North America: Apparent patterns, processes and biases. *Annual Review of Ecology and Systematics* 31:481–531.

Salen-Picard, C., and D. Arlhac. 2002. Long-term changes in a Mediterranean benthic community: Relationships between the polychaete assemblages and hydrological varia-tions of the Rhone River. *Estuaries* 25:1121–1130.

Schriever, G., A. Ahnert, H. Bluhm, C. Borowski, and H. Thiel. 1997. Results of the large scale deep-sea environmental impact study DISCOL during eight years of investi-gation. *Proceedings of the Seventh (1997) International Offshore and Polar Engineering Conference*, 1:438–444.

Sklar, F.H., and J.A. Browder. 1998. Coastal environmental impacts brought about by alterations to freshwater flow in the Gulf of Mexico. *Environmental Management* 22:547–562.

Smith, C.R., M.C. Austen, G. Boucher, C. Heip, P. Hutchings, G. King, I. Koike, J. Lambshead, and P.V.R. Snelgrove. 2000. Global change and biodiversity linkages across the sediment-water interface. *BioScience* 50:1076–1088.

Snelgrove, P.V.R., J.F. Grassle, and R.F. Petrecca. 1992. The role of food patches in maintaining high deep-sea diversity: Field experiments with hydrodynamically unbiased colonization trays. *Limnology and Oceanography* 37:1543–1550.

Somerfield, P.J., H.L. Rees, and R.M. Warwick. 1995. Interrelationships in community structure between shallow-water marine meiofauna and macrofauna in relation to dredgings disposal. *Marine Ecology Progress Series* 127:103–112.

Stachowicz, J.J., H. Fried, R.W. Osman, and R.B. Whitlatch. 2002. Biodiversity, invasion resistance, and marine ecosystem function: Reconciling pattern and process. *Ecology* 83:2575–2590.

Stevens, J.D., R. Bonfi, N.K. Dulvy, and P.A. Walker. 2000. The effects of fishing on sharks, rays and chimaeras (*chondrichthyans*) and the implications for marine ecosystems. *ICES Journal of Marine Science* 57:476–494.

Talley, T.S., and L.A Levin. 2001. Modification of sediments and macrofauna by an invasive marsh plant. *Biological Invasions* 3:51–68.

Tarazona, J., H. Salzwedel, and W. Arntz. 1988. Positive effect of "El Nino" on macrozoobenthos inhabiting hypoxic areas of the Peruvian upwelling system. *Oecologia* 76:184–190.

Teske, P.R., and T.A. Wooldridge. 2001. A comparison of the macrobenthic faunas of permanently open and temporarily open/closed South African estuaries. *Hydrobiologia* 464:227–243.

Thiel, H. 2003. Anthropogenic impacts on the deep sea. In: *Ecosystems of the World,* edited by P.A. Tyler, pp. 427–471. Amsterdam, Elsevier.

Thiel, H., G. Schriever, C. Bussau, and C. Borowski. 1993. Manganese nodule crevice fauna. *Deep-Sea Research* I 40:419–423.

Thrush, S.F., and P.K. Dayton. 2002. Disturbance to marine benthic habitats by trawling and dredging: Implications for marine biodiversity. *Annual Review of Ecology and Systematics* 33:449–473.

Thrush, S.F., D. Schultz, J.E. Hewitt, and D. Talley. 2002. Habitat structure in soft-sediment environments and abundance of juvenile snapper *Pagrus auratus*. *Marine Ecology Progress Series* 245:273–280.

Tyler, P.A. 2003. Disposal in the deep sea: Analogue of nature or faux ami? *Environmental Conservation* 30:26–39.

US Commission on Ocean Policy. 2004. Preliminary report of the US Commission on Ocean Policy—Governor's Draft. http://www.oceancommission.gov/documents/prelimreport/welcome.html. Washington, DC.

USGS Survey Fact sheet: http://marine.usgs.gov/fact-sheets/gas-hydrates/title.html

Van Dover, C.L., J.F. Grassle, B. Fry, R.H. Garritt, and V.R. Starczak. 1992. Stable isotope evidence for entry of sewage-derived organic matter into a deep-sea food web. *Nature* 360:153–156.

Vitousek, P.M., H.A. Mooney, J. Lubchenco, and J.M. Melillo. 1997. Human domination of Earth's ecosystems. *Science* 277:494–499.

Wall, D.H., P.V.R. Snelgrove, and A.P. Covich. 2001. Conservation priorities for soil and sediment invertebrates. In: *Conservation Biology,* edited by M.E. Soulé and G.H. Orians, pp. 99–123. Washington, DC, Island Press.

Wehrmann, A., M. Herlyn, F. Bungenstock, G. Hertweck, and G. Millat. 2000. The distribution gap is closed: First record of naturally settled Pacific oysters *Crassostrea gigas* in the East Frisian Wadden Sea, North Sea. *Senckenbergiana Maritima* 30:153–160.

Weston, D.P. 1990. Quantitative examination of macrobenthic community changes along an organic enrichment gradient. *Marine Ecology Progress Series* 61:233–244.

Windham, L., and R.G. Lathrop. 1999. Effects of *Phragmites australis* (common reed) invasion on aboveground biomass and soil properties in brackish tidal marsh of the Mullica River, New Jersey. *Estuaries* 22:927–935.

Worm, B., and R.A. Myers. 2004. Managing fisheries in a changing climate. *Nature* 429:15.

Appendix Table 7.A1. Vulnerability of continental shelf ecosystem services and processes to perturbations.

We summed the columns to determine which threats ranked highly among multiple services and processes, and thus contributed most to the vulnerability of each system. In cases where a range of scores was entered, the median score for that range was used in the row sums. Where information was insufficient to allow assignment of a rank value, a question mark was entered in the table and a value of zero was used in sums. Although the sums are numerically irrelevant, the relative rankings provide a yardstick of vulnerability among ecosystems and identify habitats of greatest concern.

Service	Large Scale (ocean basin)					Mid Scale (e.g., regional coast)						Small Scale (e.g., bays)			Row Sums
	Sea level	Salinity	Temperature	Wind	UV radiation	Over exploitation	Habitat loss	Persistent pollutants	Eutrophication	Hydrology change	Disease	Alien taxa	Development	Habitat loss	
Provisioning services															
Plants as food	0	1	2	2	1	1	0	1	2	1	1	2	1	1	16
Animals as food	1	−2 to 2	1	1	1	3	3	2	3	2	3	1	2	3	26
Other biological products	1	2	2	2	1	2	2	2	3	2	3	1	2	3	27
Biochemical/ medicines/models	1	1	1	1	1	1	3	1	1	1	1	?	3	3	16
Fuels/energy	0	0	0	0	0	1	?	0	0	0	0	0	0	0	1
Fiber	0	0	0	0	0	3	3	0	0	0	1	0	0	3	10
Nonliving materials (geological effects)	0	1	2	0	0	3	3	0	0	2	0	0	0	3	14
Clean seawater	0	1	1	1	−1 to 1	0	2	3	3	2	0	0	0	1	14

Regulation services															
Sediment formation: biodeposition	0	1	1	2	?	2	3	1	1	3	1	0	2	2	19
Nutrient cycling	0	1	−2 to 2	−2 to 2	1	2	3	1	3	2	0	?	1	2	16
Biological control & resistance	0	1	1	0	?	2	2	1	1	1	?	1	2	1	13
Detoxification, waste disposal	0	1	−2 to 2	−2 to 2	?	0	3	1	2	1	1	0	1	2	12
C sequestration	0	1	1	0	0	1	1	1	1	1	0	0	1	1	9
Food web support processes	0	2	−2 to 2	1	1	3	3	2	2	2	2	1	1	3	23
Atmosphere composition	0	1	−1 to 1	1	1	0	2	1	1	2	1	0	0	2	12
Flood & erosion control	0	0	−1 to 1	1	1	0	2	1	1	3	0	0	−1 to 0	3	11.5
Redox processes	0	1	−1 to 1	1	1	0	2	1	2	2	0	0	1	2	13
Habitat maintenance services															
Landscape connection & structure	0	1	1	1	0	2	3	1	1	2	1	1	−1 to +1	3	17
Aesthetic services															
Spiritual	0	0	3	2	0	3	3	1	1	2	3	0	−1	0	17
Aesthetic	0	0	0	0	0	1	1	1	1	1	1	0	1	0	7
Recreation	0	0	0	0	0	2	2	1	1	1	1	0	0	1	9
Scientific understanding	0	−1	−1	−1	−1	−1	−1	−1	−1	−1	−1	−1	−1	−1	−13
Column sums	3	15	14	15	8	31	45	22	29	32	19	6	12.5	38	289.5

Appendix Table 7.A2. Vulnerability of deep-sea ecosystem services and processes to various perturbations. See Table 7.A1 for detailed explanation.

Service	Large Scale (ocean basin)				Mid Scale (e.g., regional coast)							Small Scale (e.g., bays)			Row Sums
	Sea level	Salinity	Temperature	Wind	UV radiation	Over exploitation	Habitat loss	Persistent pollutants	Eutrophication	Hydrology change	Disease	Alien taxa	Development	Habitat loss	
Provisioning services															
Plants as food	0	0	0	0	0	0	0	0	0	0	0	0	0	0	0
Animals as food	0	0	1	1	1	3	3	1	0	1	0	?	0	2	13
Other biological products	0	0	0	0	0	1	0	0	0	0	0	0	0	1	2
Biochemical/medicines/models	0	0	0	0	0	0	0	0	0	0	0	0	0	0	0
Fuels/energy	0	0	0	0	0	2	1	0	0	0	0	0	0	0	3
Fiber	0	0	0	0	0	2	1	0	0	0	0	0	0	0	3
Nonliving materials (geological effects)	0	0	0	0	0	2	1	0	0	0	0	0	0	0	3
Clean seawater	0	0	0	0	0	0	0	0	0	0	0	0	0	0	0
Regulation services															
Sediment formation: biodeposition	0	0	1	1	0	1	1	0	0	0	0	0	0	0	4
Nutrient cycling	0	0	1	1	0	1	0	0	0	0	0	0	0	0	3

Ecosystem service	?	?	?	?	?	?	?	?	?	?	?	?	?	?	Sum
Biological control & resistance	?	?	?	?	?	?	?	?	?	?	?	?	?	?	0
Detoxification, waste disposal	0	0	1	1	0	0	0	1	1	0	0	?	0	0	4
C sequestration	0	0	3	3	0	0	0	1	0	0	0	?	0	0	7
Food web support processes	0	0	3	3	0	0	0	2	3	0	0	?	0	0	11
Atmosphere composition	0	0	1	1	0	0	0	0	0	0	0	?	0	0	2
Flood & erosion control	0	0	0	0	0	0	0	0	0	0	0	0	0	0	0
Redox processes	0	0	2	2	0	0	0	1	2	0	0	?	0	0	7
Habitat maintenance services															
Landscape connection & structure	0	0	0	0	0	0	1	1	2	0	0	?	0	0	3
Aesthetic services															
Spiritual	0	0	0	0	0	0	0	0	1	1	0	?	0	0	2
Aesthetic	0	0	0	0	0	0	0	0	0	0	0	?	0	0	0
Recreation	0	0	0	0	0	0	0	0	0	0	0	?	0	0	0
Scientific understanding	0	0	-1	-1	0	0	-1	-2	-2	-1	0	?	0	0	-7
Column sums	0	0	12	12	1	0	-1	14	15	2	1	0	0	3	60

Appendix Table 7.A3. Vulnerability of estuarine ecosystems to various perturbations. See Table 7.A1 for detailed explanation.

Service	Large Scale (ocean basin)				Mid Scale (e.g., regional coast)							Small Scale (e.g., bays)			Row Sums
	Sea level	Salinity	Temperature	Wind	UV radiation	Over exploitation	Habitat loss	Persistent pollutants	Eutrophication	Hydrology change	Disease	Alien taxa	Development	Habitat loss	
Provisioning services															
Plants as food	3	3	3	2	3	1	3	3	3	3	1	?	3	3	34
Animals as food	3	3	3	2	2	3	3	3	3	3	3	−3	2	3	33
Other biological products	3	2	2	1	0	3	3	1	2	2	1	−3	2	3	22
Biochemical/ medicines/models	1	2	1	1	1	2	3	1	1	1	?	?	2	3	21
Fuels/energy	2	0	0	0	0	2	?	0	0	0	0	0	2	2	8
Fiber	2	2	2	2	?	2	3	1	1	3	?	?	3	3	24
Nonliving materials (geological effects)	0	0	2	0	0	2	3	0	0	2	0	0	2	3	15
Clean seawater	3	3	2	3	−1	3	3	3	3	2	0	−1	2	2	27
Regulation services															
Sediment formation: biodeposition	3	2	1	3	?	2	3	1	3	3	2	−1 to 1	3	3	27
Nutrient cycling	1	1	2	1	1	2	2	3	2	2	0	?	2	2	20

															Row sums
Biological control & resistance	1	2	3	1	?	2	2	1	2	2	?	2	2	1	21
Detoxification, Waste disposal	3	2	3	2	?	?	2	1	3	2	1	1	2	2	24
C sequestration	1	1	1	0	0	1	1	1	2	1	0	0	2	2	13
Food web support processes	3	3	3	1	1	3	3	1	3	3	1	2	2	3	32
Atmosphere composition	3	2	3	2	1	1	2	1	2	2	1	2	2	3	27
Flood & erosion control	3	2	2	2	1	1	2	1	2	3	1	−3	3	3	23
Redox processes	1	1	2	1	1	1	3	1	2	3	0	2	1	2	21
Habitat maintenance services															
Landscape connection & structure	3	−2 to 2	2	2	1	2	3	1	2	3	2	−3 to 3	3	3	27
Aesthetic services															
Spiritual	?	0	3	0	0	3	3	3	3	2	3	2	3	3	28
Aesthetic	3	1	−1 to 1	0	0	1	2	1	3	2	3	−2 to 2	3	3	22
Recreation	3	1	−1 to 1	0	0	3	3	3	3	2	3	−1 to 1	−1 to 2	3	24.5
Scientific understanding	−1 to 1	−1 to 1	−1 to 1	−1 to 1	−1 to 1	−1 to 1	−1 to 1	−1 to 1	−1 to 1	−1 to 1	−1 to 1	−1 to 1	−1 to 1	−1 to 1	0
Column sums	45	33	40	26	12	40	51	32	45	46	22	0	46.5	55	493.5

PART III

Connections Between Soils and Sediments: Implications for Sustaining Ecosystems

Richard D. Bardgett

The next two chapters discuss the different types of goods and services that are provided by diverse soil and sediment biota, and highlight the vulnerability of these biota and the services they perform to global environmental change. To understand fully the roles and vulnerabilities of soil and sediment biota in a changing world requires explicit consideration of the issue of scale in terms of the spatial and temporal scale at which organism activities operate, and also consideration of the possibility that activities in one domain may affect those occurring in another. In the following chapters, both these important issues are addressed.

The first chapter addresses the issue of spatial scale of biodiversity, analyzing the scales at which biota and the services they perform are delivered, and comparing this with the scales at which the ecosystem processes operate. Different groups of the biota of soils and sediment live and operate at different spatial scales, from assemblages of microbes at the microhabitat scale to plants and macroinvertebrates that operate at much larger spatial scales. A key conclusion from this chapter is that biodiversity in soils and sediments allows the creation of self-organizing systems (SOSs) that are recognizable by a clearly defined set of interacting organisms within a specifically defined habitat, which they create and/or inhabit. Furthermore, all species within each SOS participate in providing the service to the extent that they contribute certain functional traits that can resist disturbances. In exploring this issue, the relationship between vulnerability and biodiversity is considered by detailing the functional groups and biological traits that are essential to the performance of the services and/or that allow species to resist disturbances. In view of this, it is argued that successful protection and restoration of habitats requires identification of the scales at which processes that sustain a service operate, and of the functional groups and species of biota that are essential to that service. By making these identifications, it might be possible either to limit disturbances of habitats to acceptable levels (protection) or to reintroduce, into newly recreated ecosystems, species with acceptable traits.

As noted at the start of Chapter 9 (Ineson et al.), assessment of the impacts of ecosystem management and disturbances on the provision of ecosystem services would be a comparatively simple process if ecosystems were totally self-contained and independent. This is clearly not the case: effects of disturbance or management within one spatially distinct habitat will have certain multiple effects on the biota and services of adjacent habitats. Therefore, it is essential that managers consider these secondary effects to other systems, which are an almost inevitable consequence of any adopted management policy or disturbance. Chapter 9 addresses this issue, using the example of deforestation in the terrestrial domain and demonstrating how this activity has cascading effects on the biota and services performed in adjacent, inter-connected ecosystems. The primary objective here is to strengthen our understanding of the importance of the links between the domains rather than simply emphasize the within-domain impacts. Evidence presented from the evaluation of cascading effects of deforestation suggest that changes resulting from the perturbation of a single domain will frequently be seen as impacts in other domains, and that different domains may respond in surprisingly different ways to the same environmental change—as noted in the conclusion of Chapter 9, "one domain's meat may be another domain's poison."

8

Connecting Soil and Sediment Biodiversity: The Role of Scale and Implications for Management

Patrick Lavelle, David E. Bignell, Melanie C. Austen,
Valerie K. Brown, Valerie Behan-Pelletier, James
R. Garey, Paul S. Giller, Stephen J. Hawkins, George
G. Brown, Mark St. John, H. William Hunt, and
Eldor A. Paul

Identification of the pertinent scales at which to measure various ecosystem processes, along with the recognition of possible emergent properties, are challenges that ecologists have faced over the last 20 years, often using the conceptual bases provided by hierarchy theory (Allen & Starr 1982; Meentemeyer & Box 1987; May 1989; Lavelle et al. 1993) and, more recently, landscape ecology (Urban et al. 1987; Bissonette 1997). Ecosystem services are provided across a wide range of scales and each, in turn, may stem from biota and processes that range widely in time (months, decades, centuries) or space (local, landscape, global).

The relevance of spatial scale becomes clear when we contemplate the management of ecosystems and their inherent natural biodiversity to ensure the sustainability of services under future stressors or perturbations (e.g., disease outbreaks, droughts, invasive species). Perturbations affect a range of scales; the vulnerability of each of the ecosystem services driven by biodiversity is also likely to vary, depending on the spatial scale of the perturbation and the scale on which the process or service is regulated. The degree to which biodiversity influences the resistance and resilience of ecosystem functions or processes and hence the delivery of ecological services is important in this context. We ask if this is also scale dependent.

This chapter has enormously benefited from comments and inputs by Armand Chauvel, John Mensey, Florence Dubs, Sèbastien Barot, and an anonymous referee.

As an example, in terrestrial agroecosystems, the major ecosystem service of food production occurs in crop units ranging from small plots through hectares to hundreds of kilometers. However, the heterogeneity of landscape mosaics formed by different crops, with their attendant rotation in time, is arguably the essential determinant of overall system performance. Ecological processes that underpin this production operate at a great variety of scales: from micrometers or millimeters for many basic microbial processes to meters or hundreds of meters in relation to physical soil and sediment parameters (aggregation, water infiltration), and at much larger scales if soil types or topography are concerned (see van der Putten et al., Chapter 2). Assemblages of species involved in key processes are variously distributed over these scales and their ranges of movement and dispersal vary, individually, over several orders of magnitude. Consequently, perturbations, such as local disease or pest outbreaks, overgrazing, larger-scale weather anomalies such as drought or severe winters, and longer-term climate change can have impacts across the agroecosystem, but these impacts will vary in effect depending on the magnitude, spatial extent, duration, and return rate of the perturbation. Resistance and resilience of services to press or pulse disturbances, and thus their overall vulnerability, may very much depend on the scales at which they are regulated and the scales at which the perturbations act. In many instances, management interventions will be more efficient if they operate at the scale of underlying processes rather than the scale of the service delivery.

Soils and sediments provide a range of unique ecosystem services such as nutrient cycling, carbon sequestration, and detoxification (see Wall et al., Chapter 1). These services directly or indirectly support many aboveground (or above-sediment) provisioning and cultural services (Millennium Ecosystem Assessment 2003; see also Part I, this volume). Soils and sediments are highly constraining environments where diffuse or selective mutualisms among organisms create self-organizing systems (*sensu* Perry 1995). Self-organization is inherent in soils and sediments at scales from microaggregates, created by microorganisms, to macroaggregates and the functional domains of the ecosystem engineers (Lavelle 1997); from vertical stratification profiles to soil and sediment catenas along slopes and to larger features observed at the landscape/seascape level.

A unique feature of soil and sediment systems is the range of biological activities (e.g., burrowing, tube construction, substratum translocation, and feeding processes), that sculpt and mold the substratum, generating its fine structure and heterogeneity. The diversity of ecosystem engineers is key, producing a greater variety of biogenic structures (e.g., termite mounds and galleries, earthworm casts, and burrows in soils; tubes, surface deposits, and stable aggregates; cf. ray and whale pits made by bioturbators in aquatic environments). Specific interactions among their populations are thus essential to the shaping of the aggregate and pore structure of the soil and sediment substrata. Structure, in turn, determines the ability of the substratum to accommodate and sustain microorganisms that perform specific chemical transformations, and to support other functions and processes. In soils, for example, the combined activities of "compacting" and "decompacting" invertebrates maintain macroaggregate structures, which then optimize water infiltration and retention, and limit erosion risks while sequestering carbon for both short- and long-term pools (Blanchart et al. 1999; Chauvel et al. 1999). Biogenic structures may last much longer than the ecosystem engineers that created them, especially in soils and immobile sediments. The effects of this "biological legacy" are therefore an important component of the dynamics of the engineer populations in time and space (Perry 1995).

Most management practices ignore this complexity, and by reducing biodiversity, they reduce the overall intensity of the processes mediated by biota. Managers may not recognize this cause for deterioration in the genesis of ecosystem services, since the vulnerability of ecosystems is mostly considered at the larger scale of the service delivery. For example, in the case of carbon sequestration: what happens at the scale of a cultivated field or the pelagic zone may be expressed at the scale of a watershed or a bay. To maximize our utilization of services as well as their management, assessment, and possible restoration, it is therefore essential to identify the spatial and temporal scales at which assemblages of species interact, the scales of the processes that they drive, and the resulting scales at which the various ecosystem services are delivered. It is equally important to assess the vulnerability of biodiversity and services to perturbations at all the relevant scales and the role of species/functional group richness in mediating this vulnerability.

In this chapter, we analyze the scales at which services are delivered and compare them with the scales at which the ecosystem processes operate. We suggest that the greater the difference in scale between the two, the less vulnerable the service. We then consider the relationship between vulnerability and biodiversity by detailing the functional groups and biological traits that are essential to the performance of the services (effect traits) and/or that allow species to resist disturbances (response traits). The major effect and response traits responsible for the maintenance of diversity are considered. We also propose an approach to the management of biodiversity to prevent vulnerability of ecosystem services.

Scales in Different Domains

The way we look at the habitat and the ecosystem, from the size of our sampling unit to the frequency of our observations, will influence how we identify biological responses to the environment, how we perceive the various patterns in biotic and abiotic factors, and the processes we decide are critically important in the functioning of ecosystems (Giller & Malmqvist 1998). Organizational levels in ecosystems may often be defined by a biological component, that is, the species assemblages and their complexity, and by a physical entity, that is, the structures that organisms construct and/or inhabit. Eight generalized levels can be distinguished across soil and aquatic sediment environments (see Figures 8.1a, b, and c), spanning approximately 16 orders of magnitude. At the smallest scale are assemblages of microorganisms, for example, in a microaggregate of mineral and organic material, in a micropore, or as a biofilm on a particle of substratum. Such assemblages may mediate a single fundamental process, such as carbon mineralization or nitrogen transformation. At the next level of the hierarchy are micropredators such as nematodes, which move between the consortia of microbes regulating their numbers and growth with varying degrees of feeding specialization (micro–food webs or microbial loops). At the third scale, a group of larger organisms— mainly plants (through their roots) and larger invertebrate animals—are responsible either for maintaining the physical heterogeneity of the substratum, especially mediating the mixing and re-mixing of organic and mineral material in the role of ecosystem engineers, or for fragmenting and transforming organic detritus in the role of litter transformers. These organisms, together with microorganisms, are seen as functional units in the sense that no ecosystem processes can occur unless these units are intact.

In soil systems, the individual functional units are sometimes dominated by one type of ecosystem engineer, each engineer therefore dominating its own functional domain (Lavelle 2002). Adjacent or intersecting domains make a characteristic patch or sum of patches, and these assemble to constitute a habitat. In aquatic systems, the patches tend to be more controlled by physical structures (e.g., substratum types or flow conditions) or by biotic components (e.g., macrophyte/algal stands or aggregations of sedentary or burrowing animals, litter, or organic matter accumulations). At higher levels, mosaics of functional domains and/or patches constitute habitats where the communities reside. Communities and their abiotic environment constitute an ecosystem. Mosaics (or continua) of ecosystems form the large-scale elements of ecological organization, such as catchments, landscapes, and river basins. Regions and continents, and, in the marine realm, oceans, constitute the largest spatial scales.

Clearly, total biodiversity will increase with spatial scale, as will the range of ecosystem functions and ecological services that operate, as the lower scales are nested within the larger ones. Some processes and functions are evident only at certain spatial and temporal scales, thus one cannot predict these functions and services at larger scales simply by summing those at smaller scales. There is evidence that biodiversity has a significant influence on the level and rates of ecosystem functioning and hence ecological services. Biodiversity within a single habitat patch or habitat is limited by a range of factors such as space, history of perturbations, biogeography, resource levels, and so on. Also it is evident that larger-scale species pools can influence local-scale diversity (see below). A new challenge for ecologists and managers is to understand how, as we increase spatial extent, the biodiversity of one habitat patch or habitat may influence the biodiversity and ecological functioning of adjacent patches or habitats (Giller et al. 2004).

Temporal scales reflect the rate of processes that occur across the different spatial scales (ranging from the physical and biological to the geological and evolutionary) as well as the generation time of the organisms that drive the processes. The duration and return rate of perturbations that act across the spatial gradient are also relevant. In freshwater systems, for example, small-scale surges of water flow occur at the patch level, with spates at the reach (ecosystem) level, and floods and droughts at the catchment (landscape) level and river basin scale (varying in magnitude as larger perturbations become more likely across longer time scales). Large-scale climatic events operate at the regional level. Because of the particular characteristics of the different domains, the precise spatial and temporal scales involved will differ (see Figures 8.1a, b, and c). This general scheme allows us to examine the links between scale, biodiversity, and the vulnerability of ecosystem services in all domains.

Soils

The smallest habitat in soils is represented by assemblages of mineral and organic particles approximately 20 micrometers in size, called aggregates (Figure 8.1a, level 1). Microorganisms may live inside (e.g., in micropores filled with water) or outside these constructions, which in turn determines their access to resources and exposure to predators (Hattori & Hattori 1993) and inclusion in micro–food webs (level 2; Lavelle & Spain 2001).

At the scale of centimeters to decimeters, ecosystem engineers (functional group including plant roots) and abiotic factors determine the architecture of soils through the accumulation of aggregates and pores of different sizes. These spheres of influence, or functional domains (level 3), extend horizontally over areas ranging from decimeters (e.g., the rhizosphere of a grass tussock) to 20–30 m (drilosphere of a given earthworm species) or more, and from a few centimeters up to a few meters in depth, depending on the organism. Functional domains are distributed in patches that may have discrete or nested distributions and form a mosaic of patches (level 4).

At the landscape level, different ecosystems coexist in a mosaic with clearly defined patterns (level 5). The pattern observed in the mosaic may result from natural variations in the environment and/or human land management. Soil formation processes, for example, are very sensitive to topography, which generates formation of catenas of soils from upper to lower lying areas (level 6). Significant differences in soil type at this scale often determine different vegetation types and the formation of a mosaic of ecosystems. In savanna regions of Africa, plateaus that have a thick soil and a gravel horizon are often covered with open woodland. Slopes that have more shallow soils have fewer trees; lowlying areas have fine-textured soils resulting from the transport and accumulation of fine elements from the upper-lying areas. There are also moister environments where vegetation is comprised of grasses and other herbaceous components. In riparian zones of river catchments, gallery forests may utilize constant water availability from a water table located close to the surface (Brabant 1991).

Soil formation at regional scales (level 7) is a slow process that extends over long periods of time and is largely determined by climatic conditions and the nature of the parent material. In temperate areas, for example, it takes 20,000 years to transform alu-

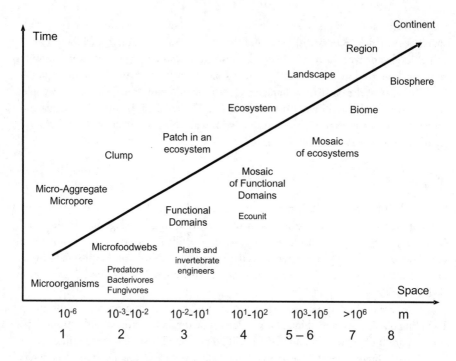

Figure 8.1a. Comparative spatial and temporal scales of biological and physical environments in soils. Levels 1–8 indicate increasing changes in the influence of biota and habitat.

mino-silicate parent material into a 1 m thick soil, but it takes half that time to develop carbonate-rich material (Chesworth 1992). Most soils in northern Europe and America that formed after the retreat of glaciers 20,000 years ago still have properties of relatively young soils, as compared with soils from Australia or some parts of Africa that began forming millions of years ago (level 8; Fyfe et al. 1983).

This hierarchical organization of soils is clearly illustrated by the determinant factors of the decomposition process (Lavelle et al. 1993). Decomposition depends on interactions between O, the organisms that operate the chemical transformations, P, the physical environment (climate and substratum), and Q, the quality of the organic matter being decomposed (Swift et al. 1979). These determinants are organized hierarchically. Decomposition is mainly mediated by microorganisms that effect over 90 percent of the chemical transformations of fresh organic residues into either mineral nutrients or recalcitrant humic acids, and operates at scales of micrometers over hours or days. At this scale, micropredators regulate and stimulate microbial activities by a well-described top-down food web effect (De Ruiter et al. 1993). In spite of this first-order regulation, microorganisms are inactive most of the time, as their movement toward fresh substratums when their local food source is depleted or limited. By contrast, macroorganisms (roots and soil invertebrates), through their bioturbation activities, stimulate microbial activities in their functional domains by a "sleeping beauty" effect (Lavelle et al. 1995) at scales of time and space of centimeters to meters over weeks to months. Interactions among micro- and macroorganisms largely depend on the quality of the organic matter produced. Low-quality organic residues, with a high content of phenolics and tannins and a low content of sugars and nutrients that can be assimilated, require more specific and efficient interactions among the biota. Thus, when litter quality is poor, macroinvertebrate activities may be greatly depressed. Coniferous or eucalyptus litters are examples of low-quality litter that result in this effect. Their decomposition is slow because only a few microbial functional groups digest the tannin-protein complexes that represent 85 percent of the nitrogen in the decomposing litter. Additionally, efficient stimulators such as earthworms or termites are mostly absent.

At higher scales of time and space, the qualities of soil and climate are the ultimate regulators of all biological activities and hence decomposition. The natural content of bedrock (nutrients and clay forming elements) determines the nutrient base and the nutrient-holding quality of clay minerals. Clay minerals provide microorganisms with numerous microhabitats, which afford suitable moisture conditions and protection from predators; however, they form barriers between microbes and organic substrates and the specific rate of microbial activity (CO_2 evolved per unit of microbial biomass) is decreased in clay compared with sandy soils (Lavelle & Spain 2001). Climate ultimately regulates biological activities through temperature and moisture conditions, at a wide range of temporal and spatial scales up to centuries and thousands to millions of kilometers, respectively.

Freshwater Sediments

Spatial scales, generally comparable to those observed in soils, are recognized in freshwater sediments (Figure 8.1b), although there is finer division at the lower scales and less spatial extensiveness at the larger scales in the soil realm than in the aquatic ones. As in the soil domain, the larger the spatial scale, the slower the processes and the rates of change. In the freshwater environment, individual sand grains and biofilms comprise the particle system (Figure 8.1b, level 1); leaf litter, rocks, other substratum patches, and macrophyte stands form the patch on a scale of centimeters (level 2), persisting for weeks to years. The habitat scale represented by level 3 includes slow-flowing pools and faster riffles and the below-surface hyporheic zone in streams and rivers. The spatial scale at these lower levels extends both horizontally and vertically within the stream substratum. River reaches (level 4) comprise linked riffles and pools forming the river habitat covering meters to tens or even hundreds of meters, persisting in their ambient locations for tens to hundreds of years. Small ponds effectively comprise the habitat scale, but larger lakes can be divided into benthic, pelagic, and littoral zones at this scale. Tributaries (level 5) merge longitudinally along catchments to form rivers that interact with

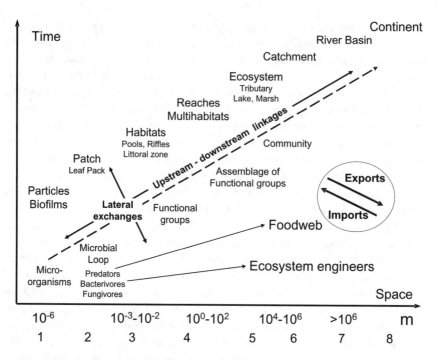

Figure 8.1b. Comparative spatial and temporal scales of biological and physical environments in freshwater sediments.

the surrounding landscape in their catchments (level 6), and whole lakes and marshes lie within the same kind of spatial and temporal scales. Lake and river chains form entire river basins (levels 7 and 8) that operate over the landscape scale and have a long geological history stretching over hundreds of thousands to millions of years. Thus, as Ward (1989) has suggested, the hierarchy of scales involved in the ecology of streams and rivers operates over four dimensions (vertical, lateral, and longitudinal spatial dimensions, and across time).

One of the most important features of this hierarchy of scales in freshwater systems is that as the spatial scale increases, the hierarchy rapidly transcends to the surrounding landscape and into groundwaters such that, in effect, at these larger spatial scales, the freshwater domain merges with the terrestrial and becomes a composite landscape ecology (Hildrew & Giller 1994). The links between freshwater systems and the surrounding landscape ensure a continuous exchange of materials, energy, and organisms (the imports and exports) between the land and the aquatic systems. Lateral exchanges and upstream-downstream linkages are thus a major feature of freshwater sediment systems (i.e., they are extremely open systems in comparison with those of terrestrial soils).

The large range of food resources available to freshwater benthic systems, from *in situ* primary production (algae and macrophytes) to organic matter produced upstream, and from the surrounding riparian zones (providing energy subsidies to the system) to other animals, has led to the evolution of a range of characteristic functional feeding groups in freshwater systems (see Giller & Malmqvist 1998 for an overview). Groups such as ecosystem engineers are common and range in scale from microengineers, such as microbes creating biofilms, to macroengineers, such as beavers influencing the structure and water flow of large river reaches. Together these assemblages of functional groups form communities that change from one habitat and ecosystem to another. Because of the openness of freshwater systems, food webs extend over a large range of spatial scales.

Freshwater systems are interesting in that, just as in soil systems, they seem to be hierarchically organized, such that the higher-scale systems impose constraints on many features of the lower scales (Frissell et al. 1986; Hildrew & Giller 1994). An example from Giller and Malmqvist (1998) illustrates this point. On a single stone, algal distribution is generally controlled by local flow and hydraulic forces, whereas in a given patch, algal biomass and species composition might be nutrient limited. Over larger stream reaches, zones of high nutrient supply can encourage algal production, allowing control to shift to grazing invertebrates or even fish, but where water chemistry controls grazers, then stream flow and substratum area are dominant. Extending the time frame, the biomass of algae in stream reaches may become determined by flood disturbance regimes and shading from riparian vegetation. Deforestation in the catchment (influencing shading, stream morphology, nutrient dynamics, and flow regimes) will influence algal dynamics, but on a very different (ecosystem) scale from the effect of grazing invertebrates. Yet such large-scale impacts in the catchment will also have implications at the small-scale level of grazer-plant interactions too, underlining the hierarchical relationships.

Hierarchical organization may extend to the regulation of the diversity of benthic organisms. Landscape factors, including size of drainage area, surrounding vegetation, geology, soils, and gradient, together with local factors such as habitat heterogeneity, system morphology, substratum and flow, water chemistry, and biotic interactions, all influence species richness in freshwater systems. In a review of benthic insect diversity in streams, Vinson and Hawkins (1998) found that richness at one scale appeared to be directly proportional to richness at higher scales, governed largely by habitat heterogeneity. For three of the major benthic insect families (Ephemeroptera [mayflies], Plecoptera [stoneflies], and Trichoptera [caddisflies]), richness at the local reach scale was 50 percent of the catchment richness, and catchment richness was 34 percent of regional richness. The regional species pool thus provides an upper theoretical limit on how many species could possibly colonize a single stream. In turn, this will limit the diversity within a reach or habitat patch (which is itself governed by local conditions).

Marine Sediments

The provision of goods and services within marine systems is dependent on biodiversity at all spatial scales, as in soils and freshwater environments (Figure 8.1c). Microbial activities underpin all ecological processes, and these in turn are affected by organismal activity at higher trophic levels and at higher levels of ecosystem organization. Functions occurring at very small scales are therefore dependent on functions at larger scales and vice-versa. Top-down regulators, however, tend to be predominant drivers expressed at larger scales, and naturally constraining processes at smaller scales. This is a reflection of the comparative openness of marine systems and also of the heterogeneity imposed on marine sediments by biota at different spatial scales.

At the smallest spatial scales of micrometers to millimeters (Figure 8.1c, level 1), microoganisms carry out the functions of decomposition, nutrient cycling, and primary production. At a slightly larger spatial scale of millimeters to centimeters, microorganisms are selectively preyed upon by meiofauna or indiscriminately consumed by selective and nonselective deposit meiofauna feeding on detritus (level 2). Microorganismal and meiofaunal biodiversity are affected by substratum properties, but these are in turn affected by physical regime. For example, coarser, well-oxygenated sediments are found in exposed high-energy environments such as wave-swept beaches and shallow coastal waters. At a scale of centimeters to hundreds of meters, larger organisms affect substratum properties (level 3). Substrata are affected at centimeter-to-meter patch sizes by individual bioturbating invertebrates (which are also usually deposit- and filter-feeding organisms), and at larger scales by groups of ecological engineers such as seagrass and mussel beds that create their own habitat. Meiofaunal diversity and community structure are also intimately connected with substratum properties and the alteration of substratum by macrofauna (Austen et al. 1998). Macrofaunal invertebrates alter substratum properties through the creation of pits, burrows, and tubes, which alter water flow at the sediment water interface, as well as through their physical movement within the sediment, and through pumping activities within tubes and burrows (level 4) (Snelgrove et al. 2000). Larger, mobile epibenthic invertebrates (e.g., decapod crustaceans) and vertebrates (e.g., pit-forming rays and whales) also similarly affect substratum properties, but pits may extend across large spatial scales of hundreds of meters and the pits themselves can be large (Feder et al 1994; Hall 1994). These activities alter water flow and diffusion of oxygen and chemicals within the sediment, which affect microbial and meiofaunal biodiversity and their functioning, and hence indirectly affect ecosystem services. These larger organisms also affect nutrient fluxes and decomposition directly through deposit and filter feeding, and through bioturbation. Nutrient cycling underpins the pelagic food web, and therefore macrofaunal invertebrates as well as microorganisms and meiofauna are important in trophic support services at larger spatial scales within the overlying water column.

Substratum-altering macrofauna often occur in localized aggregations with higher population densities (due to physical entrainment by currents during planktonic larval dispersal). For example, in soft sediment habitats there may be aggregations of burrowing shrimps, bivalve mollusks, echinoderms, or crabs creating a heterogeneous

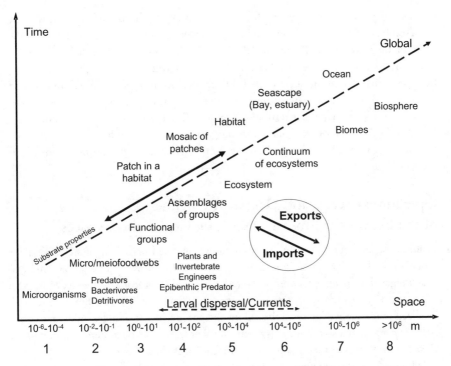

Figure 8.1c. Comparative spatial and temporal scales of biological and physical environments in marine sediments.

mosaic of overlapping patches within what would otherwise be a uniform habitat (level 5; Parry et al. 2002).

At scales of tens to hundreds of kilometers in coastal and offshore waters, different sedimentary habitats (level 6) have developed due to hydrography (the physical features of bodies of water and their littoral land areas). Fauna within these habitats are, in turn, affected by depth and distance from the coast, and the habitats can then be considered as representing inshore and offshore ecosystems. Such ecosystems occur at smaller scales within estuaries and coastal bays. Within these systems, there is input and output of organic matter to and from adjacent systems, including overlying waters. These exchanges also influence smaller substratum-dwelling organisms; thus changes in substratum inputs and outputs over large scales influence ecosystem functioning even at the smallest scales. In reality, the openness of marine systems results in a continuum of ecosystems rather than discrete systems.

At large spatial scales of meters to hundreds of kilometers, predation on macrofauna by epibenthic predators, such as crustaceans, fish, and whales, is an important component of the food web. At these scales, food provision is therefore also an important marine ecosystem service. Fishing for sediment-dwelling fauna such as shellfish and demersal fish occurs across habitats and biotopes and over spatial scales of tens to hundreds of kilometers (level 7; Jennings & Kaiser 1998; Thrush & Dayton 2002). At the largest spatial scales (level 8), benthic fauna underpin trophic support processes because of the planktonic stage in their life cycles. Vast numbers of planktonic larvae are dispersed over huge spatial scales, and these are an important component of the pelagic food web that ultimately also provides fish for human consumption. The broad-scale dispersal of macrofauna through coastal and oceanic currents also contributes to the openness of, and strong connectivity among, marine sedimentary ecosystems and habitats.

Operational Scales and Degree of Openness of the Ecosystem Process Delivering Goods and Services

Comparison of soil and aquatic domains shows that self-organizing systems regulate processes at different scales. These biological systems, however, operate at different scale ranges that increase from soils to freshwater and then from freshwater to marine ecosystems, with increasing attendant openness (connectivity). Although enclosed localized processes overlap in all three domains, marine ecosystems are much more spatially expansive than freshwater systems. By contrast, soil ecosystems are far more closed, with processes mainly operating at local scales (Figure 8.2).

The spatial scale over which ecosystem processes operate has major implications for management. In marine systems, localized management can work well for semi-enclosed ecosystems, such as lagoons or estuaries, but even these can depend on strong external linkages. Improvement of the sediment to enhance nursery ground stature for exploited fish species will work only if active spawning at a remote breeding site supplies juveniles. In contrast, soils can be manipulated at the field level or even, in traditional agrarian systems or under modern agri-environment scheme subsidies, at the level of strips within a field (e.g., establishment of species-rich headlands). For most freshwaters, the close link with the surrounding catchment means that management often has to operate at the catchment or river basin scale (see Covich et al., Chapter 3; Giller et al., Chapter 6).

The spatial scale of the process and degree of openness also determine the corresponding time scales relevant for process completion, and in turn, the effectiveness of steps to enable recovery or initiate active restoration. In marine ecosystems, the high degree of cross-connection in open systems enables recovery via propagules and larval supply from remote, unimpacted sites. The exception is when large biogenic structures are produced by ecosystem engineers (coral reefs, oyster beds, worm tubes, rooted vegetation). Thus, even in open systems active restoration has to be applied, especially as most biogenic structures proliferate via vegetative/clonal reproduction (seagrasses, coral) or where larval settlement is highly gregarious (oyster beds). Semi-enclosed sys-

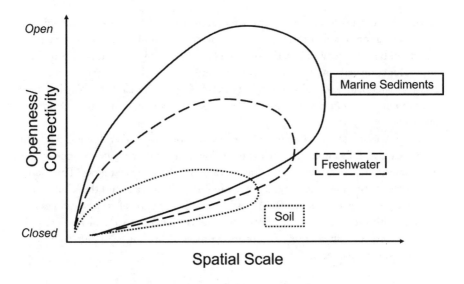

Figure 8.2. Operational scales (abscissa) and degree of openness and connectivity (ordinates) of ecosystem processes delivering goods and services in soil and aquatic sediment environments.

tems rarely recover through recruitment from unimpacted sites alone. The few examples of successful restoration or remediation in marine ecosystems come from enclosed systems in lagoons and estuaries, and usually involve a key habitat-creating feature. Similar considerations apply to freshwater systems. Although not as open as offshore water, impacted riverine systems can and do recover via recruitment from elsewhere (due to the unidirectional water flow). The short-lived flying stages of many of the aquatic insect taxa and upstream migration of aquatic stages can play a role similar to that of planktonic stages in marine life histories. Openness can, however, impede restoration attempts, as localized actions will be futile if source water is of poor quality or if disturbances in the surrounding catchment superimpose additional stresses. However, enclosed or semi-enclosed ecosystems, such as lakes, ponds, or canals, are very amenable to restoration, as a huge literature testifies (Bronmark & Hansson 1998; Giller et al., chapter 6).

By contrast, recovery of degraded soils is a slow process. Although the vegetation and some associated higher trophic levels can be actively restored by basic horticultural or agricultural practice, the soil biota and the processes they control respond much more slowly. Even if source populations are available in the surrounding landscape, the dispersal capability of most components of the soil biota is low and the timescale for population increases is correspondingly long (Brown & Gange 1990). Thus, successful restoration attempts require some form of active management of the soil in addition to management of the vegetation. Recent work (Leps et al. 2001) has shown the potential of the inoculation of soil or turf from a donor plant community to facilitate the restoration of soil biodiversity and function in ex-arable land. Even so, the time-lag in restoring soil function remains a severe constraint. Implicit in soil restoration are the problems associated with residual soil nutrient levels from previous land use. Nutrient titers are notoriously slow to change, and can therefore act as a major inhibitor of improved soil quality and function.

As the spatial scale of habitats increases, boundaries with other domains will ultimately be crossed. Examples include the transition from rivers through the riparian zone to the terrestrial landscape, and also downstream into the sea via estuaries. Thus, large-scale management must occur by manipulation of the landscape and the whole catchment (see Ineson et al., Chapter 9). A classic example is the use of afforestation as a filter along with the enhancement of other natural ecosystem services in managing the New York City watershed supply via upstream process delivering ecosystem goods (e.g., clean water; see Chapter 6).

The Vulnerability of Ecosystem Services to Biodiversity Depletion and the Impact of Scale

Soils and sediments are affected by three major kinds of stresses or perturbations that affect their biodiversity and function:

1. *Chemical pollutions* are common disturbances resulting primarily from the use of pesticides in agriculture and heavy metal contaminations. These occur in all terrestrial ecosystems, and indirectly in aquatic systems via overland transport. Eutrophication is caused by excessive transfers of nutrients from terrestrial to aquatic ecosystems. These disturbances have direct impacts on organisms and their diversity, although they do not directly affect the physical structure of the system.

2. *Physical disturbances* are caused by tillage and erosion in terrestrial systems, and by deposition or removal of sediments in aquatic ecosystems. They result in the direct destruction of the niches of organisms at scales from microhabitats to habitats or ecosystems (levels 1 to 4, Figures 8.1a, b, and c).

3. *Changes in landscapes or seascapes,* due to usage intensification or changes in uses, affect soils and sediments at level 5 to 6 whereas *global climate change* affects the largest scales (7 and 8, Figures 8.1a, b, and c). The effects of physical disturbances and intensifications on biodiversity are both direct (removal of species and habitats) and indirect (changes in fitness leading to modifications of the area of distribution).

Since disturbances may affect biodiversity at different scales, it is important to determine whether the scale at which they occur makes any difference to the resulting damage to ecosystem services, and then to identify the most appropriate scale for restoration efforts.

Are Some Ecosystem Services More Sensitive Than Others to Disturbances?

Most ecosystem services (Figures 8.3a, b, and c) may be considered as emergent properties of ecosystems that occur at the larger scales of ecosystems to landscapes/seascapes or regions. This is clearly the case, for example, for infiltration and storage of water by soils observed at the landscape level, for detoxification and decomposition in soils and sediments, and for floods and erosional events that occur at scales from the ecosystems to regional or global. On the other hand, mechanisms and processes operated by organisms and their diversity mostly occur at smaller scales, from microhabitat to habitat.

We postulate that the vulnerability of ecosystem services is dependent on the range of scales at which a disturbance impacts soil and sediment biodiversity, and similarly, is dependent on the scales at which the key biological processes operate. Our rationale is that the greater the number of scales covered by each ecosystem service and the processes that deliver them (Figures 8.3a, b, and c), the greater the sum of buffering systems represented by the different self-organizing systems operating at each level. This principle may be expressed in a number of hypotheses that were identified from a detailed analysis of these relationships.

Ecosystem services become more vulnerable as:

1. *They are delivered at smaller scales—although resilience may also be higher in that case.* This hypothesis assumes that services delivered over large scales are less likely to be severely impaired, as disturbances on a similar scale are necessary to weaken their provision. Most ecosystem services in freshwater systems (Figure 8.1b) actually operate at scales starting from 3 or 4, extending until 6 or 7, with the notable exception of nutrient cycling that operates across all levels, starting from level 1.
2. *Ecosystem services are delivered at a limited range of scales and/or supported by ecological processes operating across a small number of scales.* This hypothesis assumes that service delivery extended over a wide range of scales would provide increased buffers and enhance resistance and resilience of the provision.
3. *Biological processes that support services operate at larger scales on average.* This hypothesis assumes that if biological processes operate at small scales and the service is delivered at larger scales, disturbances in the biological components have a less immediate effect on services. This is due to the effects of biological activities, accumulated over long periods of time, which persist even after their production is interrupted with the disappearance of the organisms responsible.

Comparison of the scale ranges of services and the biological process that sustain them allows a theoretical evaluation of the vulnerability of services. Figure 8.3b illustrates freshwater ecosystems. Here, most services are operated by biological processes at smaller scales, with the exception of the habitat structure/refugia and aesthetic/recreation services. Carbon sequestration and climate regulation, on the other hand, are examples of services

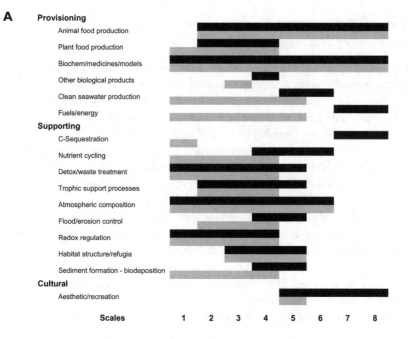

A

Provisioning
Animal food production
Plant food production
Biochem/medicines/models
Other biological products
Clean seawater production
Fuels/energy

Supporting
C-Sequestration
Nutrient cycling
Detox/waste treatment
Trophic support processes
Atmospheric composition
Flood/erosion control
Redox regulation
Habitat structure/refugia
Sediment formation - biodeposition

Cultural
Aesthetic/recreation

Scales   1   2   3   4   5   6   7   8

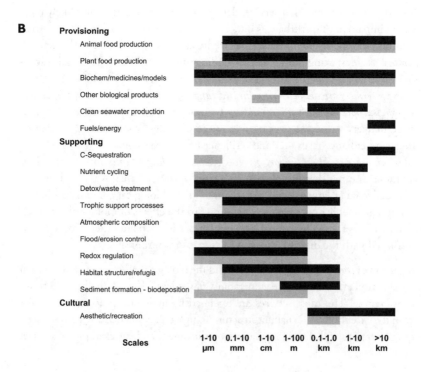

B

Provisioning
Animal food production
Plant food production
Biochem/medicines/models
Other biological products
Clean seawater production
Fuels/energy

Supporting
C-Sequestration
Nutrient cycling
Detox/waste treatment
Trophic support processes
Atmospheric composition
Flood/erosion control
Redox regulation
Habitat structure/refugia
Sediment formation - biodeposition

Cultural
Aesthetic/recreation

Scales   1-10   0.1-10   1-10   1-100   0.1-1.0   1-10   >10
           μm   mm   cm   m   km   km   km

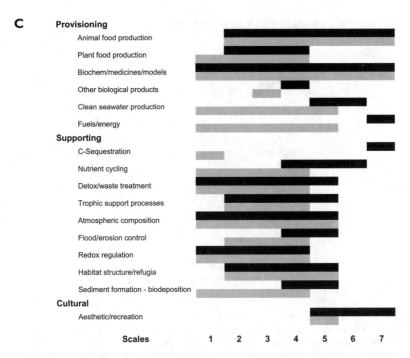

Figures 8.3a *(soil, opposite top)*, **8.3b** *(freshwater sediments, opposite bottom)*, and **8.3c** *(marine sediments, above)*. Scales at which ecosystem services are provided (black) and scales of biological processes involved (gray). Vulnerability is all the more important as the scale at which ecosystem services are delivered is restricted to a small number of levels, and biological actors operate at similar or larger scales than the ones at which ecosystem services are delivered.

delivered at much larger scales than the biological processes that operate them. These services are therefore likely to be less vulnerable than most other services, especially the habitat structure and aesthetic/recreation services that show an opposite trend.

In soils, patterns observed (Figure 8.3a) are generally similar, although the scales involved for the provision of services and processes involved tend to be smaller, as predicted by considerations of the restricted openness of the domain (Figure 8.2). Once the relevant scales for services and the corresponding processes involved are identified, optimal interventions to sustain or enhance provision of the service and/or its dependent ecological processes can be devised, at appropriate levels.

Soils

A number of soil characteristics seem to be rather resistant to disturbances. Organic matter content does not change rapidly in most manipulation experiments, and physical

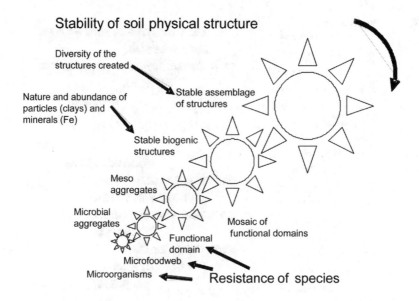

Figure 8.4. The "buffering" capacity of soils for disturbance: ecosystem services in soils are all the less vulnerable as they rely on processes operated at large scales (upper part of figure) by organisms acting at small scales (lower part of figure). This figure exemplifies the case of conservation of soil's physical structure created by the accumulation of pores and aggregates by organisms at smaller scales (smaller cogs on the lower left end), which give soils general properties (water infiltration and retention capacity) expressed at larger scales (larger cogs on the upper right end).

aggregation and porosity may be maintained in soils for some time after the organisms that have generated these features have been eliminated, for example, by mismanagement (Blanchart et al. 1999; Martius et al. 2001; Sarmiento & Bottner 2002). This resistance supports the hypothesis that lowest vulnerabilities are expected in cases where ecosystem services are supported by small-scale processes and delivered at large scales. Opposite situations would, on the other hand, lead to maximum vulnerability.

One example is soil aggregation, which is important for carbon sequestration and water infiltration, and results from accumulated activities, over long periods of time, of aggregate-forming invertebrates and roots, and the physical interactions among soil particles. The organisms operate at small spatial and temporal scales (Figure 8.1a, levels 1–3), whereas aggregates, once formed and stabilized, persist for long periods of time, depending on conditions (Blanchart et al. 1999). As a result, practices that eliminate soil ecosystem engineers may not result in an immediate impairment of soil conditions and resulting ecosystem services. Similarly, organic matter accumulated in soils is mainly the product of microbial digestions operating at the scale of individual colonies; synthesized humic compounds, on the other hand, may persist for several centuries. About half the organic matter in most soils is more than a thousand years old (Jenkinson & Rayner 1977). This particular resistance of major physical and chemical parameters give soils their capability to resist disturbances. If, however, a massive erosion affects soil at the scale of a habitat or ecosystem, services are immediately affected and natural restoration of processes at lower scales, from levels 1 to 5, will take a very long time, perhaps several to hundreds of years.

Figure 8.4 illustrates the buffering capacity of soil as a hierarchy of interlocking cogs that represent processes operating at different scales, with the physical structures created (upper part), and the elements of biodiversity that operate them (lower part).

Microbial food webs have a stability based on the diversity of the organisms they contain (De Ruiter et al. 1993). These food webs are found in soil micro- and meso-aggregates, which provide their habitats and organic resources. At the next scale up, the functional domains of ecosystem engineers support the microbial food webs by generating their own set of stable biogenic structures, which can endure beyond the life-span of the organism that created them. The ecosystem, therefore, has a diversity of organisms with the capacity to provide good soil structure. The system as a whole should have the ability to survive the temporary depletion or suppression of any one of its component organisms, but the destruction of structure—for example, by compaction or erosion—can be ameliorated only in the medium to long term by replacing the entire hierarchy of organisms and therefore the particular processes mediated by each. This may explain the time lag in many restoration activities.

Freshwater Systems

Resistance and resilience of ecological systems to environmental changes in the freshwater domain are largely related to scale. Systems and processes that operate over small scales are less resistant to disturbances (such as floods or pollution events), but may be more resilient (i.e., recover faster). Illustrating this, the vertical axis in Figure 8.1b could be considered to represent a persistence scale, with the particle and patch persisting for weeks to perhaps years, while the catchment and river basin persist for hundreds of thousands to millions of years. The magnitude and extent of the disturbance necessary to impact a system component is also related to spatial scale of the component: the larger the spatial scale, the greater the magnitude necessary to impact it (see Hildrew & Giller 1994 for further discussion). Once disturbed, however, recovery (i.e., resilience) is likely to be faster at the smaller scale (e.g., patch) than at the larger (catchment and beyond).

The most dramatic disturbance effects in freshwater systems result from continuous or directed changes in water chemistry, caused by eutrophication and acidification, particularly in lake systems. This degradation process has two origins: (1) direct influx of pollutants from sewage plants and atmospheric deposition, and (2) transfers from terrestrial ecosystems as buffer structures (riparian forests and other nutrient conservation mechanisms) are impaired and nutrient leaking occurs. Long-term changes in stream morphology and habitats through water regulation, channelization, lowering of water tables, and introduction of exotic species are other processes that contribute (Giller & Malmqvist 1998; Malmqvist & Rundle 2002). The impacts operate largely at the levels of the tributary, catchment, and entire river basin scale. Point source pollution events, on the other hand, generally occur at the reach scale and act as short-term disturbances. Because of the flowing nature of rivers, they tend to be less susceptible than lakes to suffering long-term effects from such pollution events through the dilution and self-cleaning powers of the system (this is especially the case in response to organic pollution). The self-cleaning ability of rivers is driven to a large extent by the presence of benthic species and functional groups (such as microbes and macroinvertebrate detritivores) that break down organic matter. There is some evidence of a positive link between diversity and decomposition rates in rivers (Jonsson & Malmqvist 2000), and certainly for the presence of specific (key) species mediating the resilience (Jonsson et al 2002; see also Chapters 3 and 6, this volume). The use of wetlands as natural water treatment systems also testifies to this self-cleaning ability (Reed et al 1995; Ewel 1997). However, very extensive, concentrated, and/or continuous inputs of pollutants overcome even this innate resistance. Pollutants tend to accumulate in lakes, particularly in the sediments, and they show lower self-cleaning powers than other freshwater systems. The effects of various disturbances on ecosystem services in freshwater systems are considered in detail in Chapter 6.

Recovery following disturbances varies with the nature, magnitude, and extent of the disturbance, but on the whole freshwater systems are quite resilient, and most studies indicate fairly rapid recovery, at least in river systems (Niemi et al. 1990). However, bioaccumulation and biomagnification of the more highly persistent pollutants (such as heavy metals) are well documented in freshwater systems (particularly in lakes and large rivers), and thus pose a longer-term and more serious threat to ecosystem services.

Marine Systems

The vulnerability of marine systems at different spatial scales is dealt with, in detail, by Snelgrove et al. (Chapter 7). Scale-dependent vulnerability varies with habitat (estuarine, offshore, deep-sea, cave and sub-sedimentary) and proximity to anthropogenic threat. Coastal and estuarine systems are more vulnerable than the deep seas. As with freshwater systems, resistance and resilience in the marine domain are largely related to scale. Smaller, contained systems such as estuaries and sub-sedimentary habitats are more vulnerable to disturbance but they are also more resilient; functions can be partially restored through active restoration projects and passive recruitment processes (Hawkins et al. 1999, 2002). The progressively increasing openness and connectivity of estuarine, offshore, and deep-sea systems allows for a degree of resilience at all scales, as most species can potentially recruit widely dispersed propagules or larvae from remote sources, and succession is usually rapid. Equally, the openness of most marine systems increases resistance to changes in water quality at medium to large scales due to dilution effects, but decreases resilience at larger scales where there is extensive and continuous input leading to eutrophication and the accumulation of toxic material, particularly in sediments. Vulnerability of marine systems varies with the scale of threat.

Habitat loss occurs at small, localized scales often within coastal and estuarine areas through land reclamation for coastal development. This can scale up to become whole coastline effects (e.g., the coast of Europe from Belgium to Denmark). Larger-scale disruption of sediments can be caused by fishing (trawling) activities over large areas of the continental shelf. All habitats at all scales are affected by climate change, but there is a degree of resistance to larger-scale temperature changes as these are buffered by the water, particularly in offshore and deep-sea systems. Changes in temperature, wind, and weather patterns, as well as sea-level rise, have more impact in coastal and estuarine waters, particularly in intertidal and shallow water coastal margins, where there is less buffering capacity in the overlying water.

Functional Groups Responsible for Ecosystem Services and Their Vulnerability

Previously cited examples have shown the contribution of the associations of soil and sediment organisms and plants to the creation of structures and other buffering mechanisms at different scales, providing resistance and resilience for processes delivering ecosystem services. These self-organizing systems depend on a number of species that make different contributions in the provision of the service, and also have different responses to disturbances. Thus, it is essential to predict the effect of disturbances on the patterns of disassembly of such communities and the consequences for the delivery of services. A similar approach is also required to devise an efficient restoration of these communities and their associated buffering systems.

To manage ecological services effectively and to understand the relationship between vulnerability of the service and biodiversity, we need a tool to evaluate the role of different species in providing resilience and resistance to perturbations. It is clear that not all species have the same impacts, nor do they present equivalent resistance or resilience capabilities when disturbances occur. We distinguish functional groups that have biological traits necessary to the performance of a given function (effect traits) from response groups of species that may resist stress or disturbance by a number of response traits (Lavorel et al. 1997; Diaz & Cabido 2001). The concept of functional groupings offers a means of classifying organisms according to their contribution to these processes.

Table 8.1. Major functional groups of organisms found in soil and sediment environments.

Functional Group Number	General Function	Domain— Responsible Functional Groups		
		Soil	Marine	Fresh Water
1	Decomposers	Microbes	Microbes, meiofauna	Biofilm producers, microbial respirers
2	Mutualists/ symbionts	Microbes	Microbes	
3	N$_2$-Fixers	Microbes	Microbes	Microbes
4	S-transformers	Microbes	Microbes	Microbes
5	Microengineers	Microbes & invertebrates	Microbes & invertebrates	Biofilm producers, filterers
6	Trace gas producers/ removers	Microbes	Microbes	Microbes
7	Primary producers	Algae, microbes, plants	Micro- and macroalgae, vascular plants	Algae macrophytes
8	Herbivores	Invertebrate grazers	Invertebrate grazers	Invertebrate grazers
9	Detritivores— litter transformer/ shredder	Invertebrates	N/A	Invertebrates
10	Detritivores— deposit feeder	N/A	Invertebrates	Invertebrates
11	Detritivores— filter feeders	N/A	Filter feeders	Filterers
12	Predators	Invertebrates and vertebrates	Epibenthic, infauna	Invertebrates and vertebrates
13	Pathogens	Microbes, viruses, prions	Microbes, viruses, prions	Microbes, viruses, prions
14	Bioturbators	Invertebrates	Invertebrate and vertebrate	Invertebrates
15	Macroengineers	Invertebrates and vertebrates	Invertebrates and vertebrates	Invertebrates

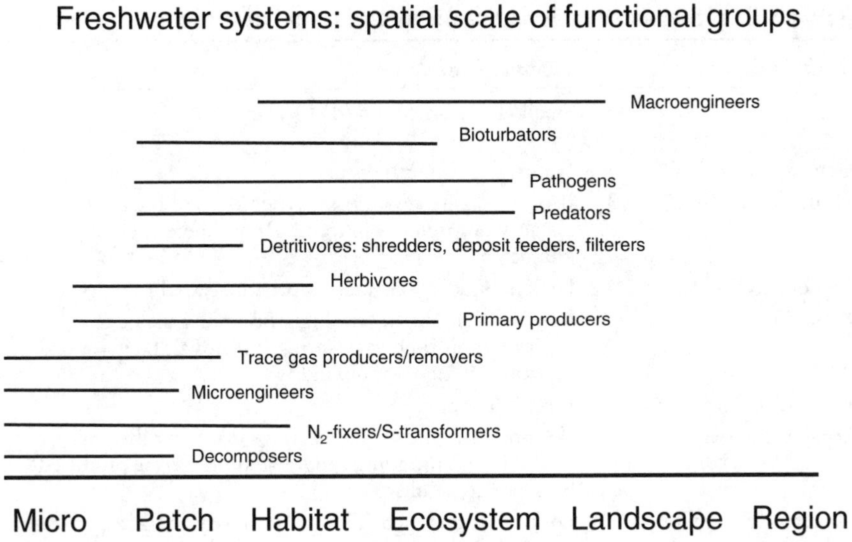

Freshwater systems: spatial scale of functional groups

```
                    ─────────────────────────────────  Macroengineers
        ─────────────────────────────────  Bioturbators
        ─────────────────────────────────  Pathogens
        ─────────────────────────────────  Predators
            ──────────  Detritivores: shredders, deposit feeders, filterers
    ──────────────────────  Herbivores
    ─────────────────────────────────  Primary producers
──────────────────────  Trace gas producers/removers
──────────────  Microengineers
──────────────────────  N₂-fixers/S-transformers
──────────────  Decomposers
```

| Micro | Patch | Habitat | Ecosystem | Landscape | Region |

Figure 8.5. Spatial scales of functional groups in freshwater systems.

Furthermore, although terminologies vary between domains, there is a degree of consistency in function of the groupings.

A set of 15 common functional groups may occur in soil, freshwater, and/or sediment environments, representing a wide variety of major taxonomic groups (Table 8.1). Each functional group includes a variable number of species, from one species to relatively species-rich phylogenetic clades (Brussaard et al. 1997). Response traits of the species that comprise these functional groups determine the overall vulnerability of a functional group to perturbation, and also affect the scale of this vulnerability. For example, species with short generation times, high fecundity, and high growth rates are often ecologically tolerant and widely distributed, and they are, characteristically, successful colonists (Sakai et al. 2001).

As an example of an operational spatial scale that is midway between that of soil and marine sediments, the spatial scale of occurrence of these functional groups may be mapped from the microscopic level to landscape to region in freshwater (Figure 8.5). Figure 8.5 shows the scales of operation for our hypothesized set of generic functions. The range is related to the physical system components (levels 1 to 8) given in Figure 8.1b. The macroengineers, for example, can influence the system up to the landscape scale, as clearly illustrated by the beaver's activity in damming streams and changing downstream and lateral flow patterns and movement of materials. Bioturbators can have a small patch/habitat scale effect, as in the case of salmonid reproductive behav-

Table 8.2. Life history traits: Definition and scope of categories.

Trait	States Possible
Size	Small (1–3)*, medium (4–6)*, large (7–8)*.
Life span/ Generation interval**	Hours, days, months, years.
Dispersal capacity	Poor, moderate, good. Assessed in terms of the typical distance propagules can travel. Poor = meters, good = kilometers.
Reproductive mode	Fission, budding, asexual propagules, sexual gametes.
Fecundity	Low, medium, high, very high. Fecundity includes offspring/unit time or offspring/individual. Low = single figures yr^{-1}, high = thousands yr^{-1}.
Facultative dormancy	Present, absent.
Separate life stages	Yes, no. Examples are insect larvae living in different habitats from the adults, and planktonic larvae produced by sedentary adults.
Ecological flexibility	Low, moderate, high. This is a summary character, reflecting the ability of an organism to switch substrates or habitats in response to disturbances or stresses.

*Numbers refer to spatial scale categories in Figures 8.1a–8.1c.
**These would not be the same if there were dormancy.

ior, which produce redds to lay their eggs in the substratum. In addition, larger-scale effects can occur, such as those of muskrats or hippos in wetlands. The openness of the freshwater systems allows predators and pathogens to operate over larger scales than might be expected from their size, whereas detritivores tend to operate at the patch to habitat level, associated with the detritus patches. Although primary producers in streams are at fairly small scales, phytoplankton in large lakes function at the full ecosystem scale. Microengineering is illustrated in freshwater systems by algal mats in desert streams, microbes/biofilm stabilizing fine sediment, and even by blackfly larvae that filter fine particles and produce larger-sized fecal particles, which leads to changes in particle size distribution and influences the movement of fine particulate matter downstream (Giller & Malmqvist 1998). Microbial decomposer activity is generally on small scales, although nitrogen fixation by blue-green algal mats can occur at the habitat scale.

Biological response traits of soil, freshwater, and marine species can be classified in eight types: size, potential life span, voltinism, dispersal capability, reproductive mode (e.g., sexual, asexual, facultatively asexual), fecundity, phenotypic plasticity (such as facultative dormancy and occurrence of spatially separated life stages), and ecological flexibility (or tolerance/amplitude of response). These response traits can be scaled against the corresponding functional groups (Table 8.2). One handicap is that life-history traits are poorly known for most species, other than some vertebrates, higher plants, and some invertebrates, (e.g., Brown 1986 for insects; Grime et al. 1988 for plants; Lavelle & Spain 2001 for soil invertebrates). Furthermore, phenotypic plasticity in many populations (such as facultative dormancy and occurrence of spatially separated life stages) and the range of triggers for this plasticity seem to be more extensive than previously expected (Lavelle 1983; Negovetic & Jokela 2001; Jennions & Telford 2002; Mitchell & Carvalho 2002; Peckarsky et al. 2002). This is particularly relevant to the delivery of ecosystem services. For example, phenotypic plasticity is a biological characteristic of successful invaders, just as invasive ant species may be characterized by the formation of supercolonies that lack distinct behavioral boundaries and are, therefore, able to dominate entire habitats (Holway et al. 2002).

Current risk assessment protocols for weeds, invasive plants, and alien fish species in North America are predicated on knowing species characteristics (e.g., Kolar & Lodge 2002), as are conservation efforts (Côte & Reynolds 2002), though similar approaches to soil and sediment organisms are limited to very few cases (Blackshaw 1997; McLean & Parkinson 2000). Recently, Wilby and Thomas (2002) suggested that basic biological insights can "allow more accurate predictions of the effect of species loss on the delivery of ecosystem services," but these are available only for a limited number of soil or aquatic organisms (Lavelle & Spain 2001). Several taxa have a wide range of biological traits, such that the functional group of "litter transformers" can include species with high metabolism, high fecundity, and short life spans (e.g., most microarthropods including Astigmata), and those with low fecundity and long life spans, most taking more than a year to complete their life cycle (e.g., most Oribatida; Walter & Proctor 1999). Insect root herbivores are, likewise, variable in life-history traits and may display variation according to abiotic and biotic conditions (Brown & Gange 1990). For microorganisms, there has been little attention to life-history traits, other than reproductive attributes. More information is available on soil ecosystem engineers and their functional impact on soil processes. Recently, attention has been paid to the structure and composition of the biogenic structures that they produce, and their evolution in time and impact on soil properties at large scales (Le Bayon & Binet 1999; Decaëns et al. 2001). Specific microbial communities generate characteristic organic signatures (based on near infrared spectrometry), which have been characterized to indicate the likely diversity of functional impacts in soil (Hedde, Decaëns, Jimenez, and Lavelle, unpub. data).

Table 8.3 presents our assessment of the most critical set of ecosystem services, listed, respectively, as provisioning, supporting, and cultural services, and the functional groups that are responsible for delivery of each. For the listed services, traits contributing to vulnerability of the service and the desired trait level can be identified. An example is given in Table 8.4 for nutrient cycling in freshwater systems.

Table 8.3. Functional groups involved in the provision of ecosystem services.

Service	Freshwater	Marine	Soil
		Functional Group Involved	
Provisioning			
Animal food production	8,12,11	8,10,11,12,14,15	none
Plant food production	7,8	7,8	1,7
Other biological production	none	8,10,11,12,14,15	none
Biochemicals/medicines/models	7	1,2,6,7,8,10,11	1,7,15
Fresh water production	1,9,10,4		1,15
Clean sea water	N/A	1,4,5,6,7,8,10, 11,13,14,15	N/A
Fuels/energy	7	1,3,5,7,8,10, 11,12,14,15	N/A
Nonliving material		1,3,4,7,8,15	1,3,7
Transport	7,15	N/A	N/A
Supporting			
C sequestration	1,7	1,7	1,5,7,15
Trace gasses/Atmospheric composition	6		6,15
Soil and sediment formation + structure	5,14,15	1,5,7,8,10, 11,12,14,15	1,5,7,14,15
Nutrient cycling	1,3,6,9,10,11	1,5,7,8,10, 11,12,14,15	1,2,3,4,6,7,9, 12,14,15
Biocontrol			
Detoxification/waste treatment	1,6,9,10,11	1,3,4,5,6,7,8,10, 11,13,14,15	1,6,7
Flood/erosion control	7,15	1,5,7,8,10,11,14,15	5,7,14,15
Climate regulation/ Atmospheric composition	3,4,6	1,3,4,5,6,7,8,10, 11,12,13,14,15	1,2,3,4,6,14,15
Redox	1,6	1,2,3,4,6,7,8,10, 11,13,14,15	N/A
Trophic support		1,2,7,8,10,11, 12,13,14,15	
Habitat provision/refugia/landscape interconnections + structure	5,7,14,15	1,2,5,7,8,10, 11,12,13,14,15	5,7,14,15
Cultural			
Aesthetic/recreation	7,8,12	1,3,4,5,6,7,8,10, 11,12,13,14,15	1,14,15

Table 8.4. Nutrient cycling in freshwater systems: Functional groups and associated biological traits responsible for the service, scales at which the service is delivered, and optimum intervention levels for management along biotic and spatial scales as defined in Figures 8.1a–8.1c.

Spatial		Biotic			Traits	Desired
	Intervention		Intervention	Functional	Contributing	Trait
Scale	Level	Scale	Level	Group	to Vulnerability	Level
1–6	5–6	1–4	4	1,3,6	DC	7
					EF	Generalist
				9,10,11	LS	5–6
					DC	6–7
					PP	No
					EF	Generalist
					F	4–6

Implications for Management of Ecosystem Services

Based on the eight general levels for spatial and biotic patterns we identified earlier (Figures 8.1a, b, and c), we have been able to consider the scales over which the various ecosystem services are most vulnerable, at what scale we can intervene for management, and which functional groups are the most important for the various services. We can then evaluate which of the response traits (Table 8.2) of species are contributing most to their vulnerability, and what the desired trait level would be to overcome/resist this vulnerability. While this is currently a conceptual exercise, it does offer the possibility for the development of a management tool once we have more objective data on specific vulnerability levels and critical traits. As an example of the approach of identifying biological/ecological traits that contribute to the vulnerability of a particular service for the freshwater domain, we have used the ecosystem service of nutrient cycling. The desired trait levels are "best guesses" at present, until research has rigorously tested the links between traits, the functioning of the ecosystem, and the provision and stability of the services. The approach offers a good basis for the development of better management practices and a method for linking considerations of biodiversity more directly to the management of ecological services.

We conclude with a number of clear patterns that emerge from this analysis. We suggest that these patterns provide a useful basis for considering the appropriate scales of management intervention in soils and sediments, with a focus on the key biodiversity elements involved:

1. The biodiversity in soils and sediments allows the creation of self-organizing systems (SOSs), recognizable by a clearly defined set of interacting organisms and a specifically defined habitat that they create and/or inhabit. SOSs are nested within each other as scale increases.
2. SOSs participate in the regulation and/or delivery of ecosystem services at eight different levels extending over approximately 16 orders of magnitude, from microhabitats to the whole biosphere. Each of these SOSs has a definable buffering capacity against disturbances.
3. All species in SOSs participate in provision of the service to the extent that they contribute certain functional traits and can resist disturbances with adequate response traits.
4. Ecosystem services provided by soils and sediments are more vulnerable as they are provided at a limited number of scales (with a low number of buffers), and because a limited number of scales separate the processes from their delivery.
5. Successful protection and restoration therefore requires a number of specific steps:
   - Identify the scales at which processes that sustain a service operate;
   - Identify the functional groups that are essential to the service and the species that comprise the functional group;
   - List the response traits that allow these species to adapt to disturbance;
   - Limit disturbances to acceptable levels (protection) or reintroduce species with acceptable trait levels in a newly recreated ecosystem.

Literature Cited

Allen, T.F.H., and T.B. Starr. 1982. *Hierarchy. Perspectives for Ecological Complexity.* Chicago, Illinois: The University of Chicago Press.

Austen, M.C., S. Widdicombe, and N. Villano-Pitacco. 1998. Effects of biological disturbance on diversity and structure of meiobenthic nematode communities. *Marine Ecology Progress Series* 174:233–246.

Bissonette, J.A. 1997. Scale-sensitive ecological properties: Historical context, current meaning. In: *Wildlife and Landscape Ecology: Effects of Pattern and Scale,* edited by J.A. Bissonette, pp. 3–31. New York, Springer-Verlag.

Blackshaw, R.P. 1997. Life cycle of the earthworm predator *Artioposthia triangulata* (Dendy) in Northern Ireland. *Soil Biology and Biochemistry* 29:245–249.

Blanchart, E., A. Albrecht, J. Alegre, A. Duboisset, B. Pashanasi, P. Lavelle, and L. Brussaard. 1999. Effects of earthworms on soil structure and physical properties. In: *Earthworm Management in Tropical Agroecosystems,* edited by P. Lavelle, L. Brussaard, and P. Hendrix, pp. 139–162. Wallingford, UK, CAB International.

Brabant, P. 1991. *Le sol des forêts claires du Cameroun.* Paris, ORSTOM-MESIRES.

Bronmark, C., and L.-A. Hansson. 1998. *The Biology of Lakes and Ponds.* Oxford, Oxford University Press.

Brown, V.K. 1986. Life cycle strategies and plant succession. In: *The Evolution of Insect Life Cycles,* edited by F. Taylor and R. Karban, pp. 105–124. New York, Springer-Verlag.

Brown, V.K., and A.C. Gange. 1990. Insect herbivory below ground. *Advances in Ecological Research* 20:1–58.

Brussaard, L., V.M. Behan-Pelletier, D.E. Bignell, V.K. Brown, W. Didden, P. Folgarait, C. Fragoso, D.W. Freckman, V.V.S.R. Gupta, T. Hattori, D.L. Hawksworth, C. Klopatek, P. Lavelle, D.W. Malloch, J. Rusek, B. Söderström, J.M. Tiedje, and R.A. Virginia. 1997. Biodiversity and ecosystem functioning in soil. *Ambio* 26:563–570.

Chauvel, A., M. Grimaldi, E. Barros, E. Blanchart, T. Desjardins, M. Sarrazin, and P. Lavelle. 1999. Pasture degradation by an Amazonian earthworm. *Nature* 389:32–33.

Chesworth, W. 1992. Weathering systems. In: *Weathering, Soils, and Palaeosols,* edited by I.P. Martini and W. Chesworth, pp. 19–40. Amsterdam, Elsevier.

Côté, I.M., and J.D. Reynolds. 2002. Predictive ecology to the rescue? *Science* 298:1181–1182.

Decaëns, T., J.H. Galvis, and E. Amezquita. 2001. Properties of the structures created by ecological engineers at the soil surface of a Colombian savanna. *Comptes Rendus De l' Académie des Sciences, Série III,* 324:465–478.

De Ruiter, P.C., J.A. van Veen, J.C. Moore, L. Brussaard, and H.W. Hunt. 1993. Calculation of nitrogen mineralization in soil food webs. *Plant Soil* 157:263–273.

Diaz, S., and M. Cabido. 2001. Vive la différence: Plant functional diversity matters to ecosystem processes. *Trees in Ecology and Evolution* 16:646–655.

Ewel, K.C. 1997. Water quality improvement by wetlands. In: *Nature's Services: Societal Dependence on Natural Ecosystems,* edited by G.C. Daily, pp. 329–344. Washington, DC, Island Press.

Feder, H.M., J.S. Naidu, S.C. Jewett, J.M. Hameedi, W.R. Johnson, and T.E. Whitledge. 1994. The Northeastern Chukchi Sea: Benthos-environmental interactions. *Marine Ecology Progress Series* 111:171–190.

Frissell, C.A., W.J. Liss, C.E. Warren, and M.D. Hurley. 1986. A hierarchical framework for stream habitat classification: Viewing streams in a watershed context. *Environmental Management* 10:199–214.

Fyfe, W.S., B.I. Kronberg, O.H. Leonardos, and N. Olorunfemi. 1983. Global tectonics and agriculture: A geochemical perspective. *Agriculture, Ecosystems and Environment* 9:383–399.

Giller, P.S., H. Hillebrand, U. Berninger, M.O. Gessner, S.J. Hawkins, P. Inchausti, C. Inglis, H. Leslie, B. Malmqvist, M.T. Monaghan, P.J. Morin, and G. O'Mullan. 2004. Biodiversity effects on ecosystem functioning: Emerging issues and their experimental test in aquatic environments. *Oikos* 104:423–436.

Giller, P.S., and B. Malmqvist. 1998. *The Biology of Streams and Rivers.* Oxford, Oxford University Press.

Grime, J.P., J.G. Hodgson, and R. Hunt. 1988. *Comparative Plant Ecology: A Functional Approach to Common British Species.* London, Unwin Hyman.

Hall, S.J. 1994. Physical disturbance and marine benthic communities: Life in unconsolidated sediments. *Oceanography and Marine Biology and Annual Review* 32:179–239.

Hattori, R., and T. Hattori. 1993. Soil aggregates as microcosms of bacteria-protozoa biota. *Geoderma* 56:493–501.

Hawkins, S.J., J.R. Allen, and S. Bray. 1999. Restoration of temperate marine and coastal ecosystems: Nudging nature. *Aquatic Conservation-Marine and Freshwater Ecosystems* 9:23–46.

Hawkins, S.J., J.R. Allen, P.M. Ross, and M.J. Genner. 2002. Marine and coastal ecosystems. In: *Handbook of Ecological Restoration. Volume 2, Restoration in Practice*, edited by M.R. Perrow and A.J. Davy, pp. 121–148. Cambridge, UK, Cambridge University Press.

Hildrew, A.G., and P.S. Giller. 1994. Patchiness, species interactions and disturbance in the stream benthos. In: *Aquatic Ecology: Scale, Pattern and Process*, edited by P. Giller, A. Hildrew, and D. Raffaelli, pp. 21–62. Oxford, Blackwell Scientific.

Holway, D.A., L. Lach, A.V. Suarez, D.V. Tsutsui, and T.J. Case. 2002. The causes and consequences of ant invasions. *Annual Review of Ecology and Systematics* 33:181–233.

Jenkinson, D.S, and J.H. Rayner. 1977. The turnover of soil organic matter in some of the Rothamsted classical experiments. *Soil Science* 123:298–305.

Jennings, S., and M.J. Kaiser. 1998. The effects of fishing on marine ecosystems. *Advances in Marine Biology* 34:203–314.

Jennions, M.D., and S.R. Telford. 2002. Life-history phenotypes in populations of *Brachyrhaphis episcopi* (Poeciliidae) with different predator communities. *Oecologia (Berlin)* 132:44–50.

Jonsson, M., O. Dangles, B. Malmqvist, and F. Guerold. 2002. Simulating species loss following a perturbation: Assessing the effects on process rates. *Proceedings of the Royal Society of London, Series B—Biological Sciences* 269:1047–1052.

Jonsson, M., and B. Malmqvist. 2000. Ecosystem process rate increases with animal species richness: Evidence from leaf-eating, aquatic species. *Oikos* 89:519–523.

Kolar, C.S., and D.M. Lodge. 2002. Ecological predictions and risk assessment for alien fishes in North America. *Science* 298:1233–1236.

Lavelle, P. 1983. The structure of earthworm communities. In: *Earthworm Ecology: From Darwin to Vermiculture*, edited by J.E. Satchell, pp. 449–466. London, Chapman & Hall.

Lavelle, P. 1997. Faunal activities and soil processes: Adaptive strategies that determine ecosystem function. *Advances in Ecological Research* 27:93–132.

Lavelle, P. 2002. Functional domains in soils. *Ecological Research* 17:441–450.

Lavelle, P., E. Blanchart, A. Martin, S. Martin, I. Barois, F. Toutain, A. Spain, and R. Schaefer. 1993. A hierarchical model for decomposition in terrestrial ecosystems: Application to soils in the humid tropics. *Biotropica* 25:130–150.

Lavelle P., C. Lattaud, D. Trigo, and J.I. Barois, 1995. Mutualism and biodiversity in soil. *Plant and Soil* 170:23–33.

Lavelle, P., and A.V. Spain. 2001. *Soil Ecology*. Amsterdam: Kluwer Scientific Publications.

Lavorel, S., S. McIntyre, J. Landsberg, and T. Forbes. 1997. Plant functional classifications: From general functional groups to specific groups of response to disturbance. *Trends in Ecology and Evolution* 12:474–478.

Le Bayon, R.C., and F. Binet. 1999. Rainfall effects on erosion of earthworm casts and phosphorus transfers by water runoff. *Biology and Fertility of Soil* 30:7–13.

Leps, J., V.K. Brown, T.A. Diaz Len, D. Gormsen, K. Hedlund, J. Kailova, G.W. Korthals, S.R. Mortimer, C. Rodriguez-Barrueco, J. Roy, I. Santa Regina, C. van Dijk, and W.H. van der Putten. 2001. Separating the chance effect from other diversity effects in the functioning of plant communities. *Oikos* 92:123–134.

Malmqvist, B., and S. Rundle. 2002. Threats to the running water ecosystems of the world. *Environmental Conservation* 29:134–153.

Martius, C., H. Tiessen, and P.L.G. Vlek, editors. 2001. *Managing Organic Matter in Tropical Soils: Scope and Limitations*. Dordrecht, The Netherlands, Kluwer Academic Publishers.

May, R.M. 1989. Levels of organization in ecology. In: *Ecological Concepts: The Contribution of Ecology to an Understanding of the Natural World*, edited by J.M. Cherret, pp. 339–363. New York, Blackwell Science Publishers.

McLean, M.A., and D. Parkinson. 2000. Field evidence of the effects of the epigeic earthworm *Dendrobaena octaedra* on the microfungal community in pine forest floor. *Soil Biology and Biochemistry* 32:351–360.

Meentemeyer, V., and E.O. Box. 1987. Scale effects in landscape studies. In: *The Role of Landscape Heterogeneity in the Spread of Disturbance,* edited by M.G. Turner, pp. 13–34. New York, Springer Verlag.

Millennium Ecosystem Assessment. 2003. *Ecosystems and Human Well-Being: A Framework for Assessment.* Washington, DC, Island Press.

Mitchell, S.E., and G.R. Carvalho. 2002. Comparative demographic impacts of "infochemicals" and exploitative competition: An empirical test using *Daphnia magna. Freshwater Biology* 47:459–471.

Negovetic, S., and J. Jokela. 2001. Life-history variation, phenotypic plasticity, and subpopulation structure in a freshwater snail. *Ecology* 82:2805–2815.

Niemi, G.J., P. Devore, N. Detenbeck, D. Taylor, A. Lima, J. Pastor, J.D Yount, and R.J. Naiman. 1990. Overview of case studies on recovery of aquatic systems from disturbance. *Environmental Management* 14:571–587.

Parry, D.M., L.A. Nickell, M.A. Kendall, M.T. Burrows, D.A. Pilgrim, and M.B. Jones. 2002. Comparison of abundance and spatial distribution of burrowing megafauna from diver and remotely operated vehicle observations. *Marine Ecology Progress Series* 244:89–93.

Peckarsky, B.L., A.R. McIntosh, B.W. Taylor, and J. Dahl. 2002. Predator chemicals induce changes in mayfly life history traits: A whole-stream manipulation. *Ecology* 83:612–618.

Perry, D.A. 1995. Self-organizing systems across scales. *Trends in Ecology and Evolution* 10:241–245.

Reed, S.C., R.W. Crites, and E.J. Middlebrooks. 1995. *Natural Systems for Waste Management and Treatment,* 2nd edition. New York, McGraw-Hill Inc.

Sakai, A.K., F.W. Allendorf, J.S. Holt, D.M. Lodge, J. Molofsky, K.A. With, S. Baughman, R.J. Cabin, J.E. Cohen, N.C. Ellstrand, D.E. McCauley, P. O'Neil, I.M. Parker, J.M. Thompson, and S.G. Weller. 2001. The population biology of invasive species. *Annual Review of Ecology and Systematics* 32:305–332.

Sarmiento, L., and P. Bottner. 2002. Carbon and nitrogen dynamics in soils with different fallow times in the high tropical Andes: Indications for fertility restoration. *Applied Soil Ecology* 19:79–89.

Smith, C.R., M.C. Austen, G. Boucher, C. Heip, P.A. Hutchings, G.M. King, I. Koike, P.J.D. Lambshead, and P. Snelgrove. 2000. Global change and biodiversity linkages across the sediment-water interface. *BioScience* 50:1108–1120.

Snelgrove, P.V.R., M.C. Austen, G. Boucher, C. Heip, P.A. Hutchings, G.M. King, I. Koike, P.J.D. Lambshead, and C.R. Smith. 2000. Linking biodiversity above and below the marine sediment-water interface. *BioScience* 50:1076–1088.

Swift, M.J., O.W. Heal, and J.M. Anderson. 1979. *Decomposition in Terrestrial Ecosystems.* Oxford, Blackwell Scientific Publications.

Thrush, S.F., and P.K. Dayton. 2002. Disturbance to marine benthic habitats by trawling and dredging: Implications for marine biodiversity. *Annual Review of Ecology and Systematics* 33:449–473.

Urban, D.L., R.V. O'Neill, and H.H. Shugart Jr. 1987. Landscape ecology: A hierarchical perspective can help scientists understand spatial patterns. *BioScience* 37:119–127.

Vinson, M.R., and C.P. Hawkins. 1998. Biodiversity of stream insects: Variation at local, basin, and regional scales. *Annual Review of Entomology* 43:271–293.

Walter, D.E., and H.C. Proctor. 1999. *Mites: Ecology, Evolution and Behaviour*. Sydney, University of New South Wales Press, and Wallingford, UK, CABI.

Ward, J.V. 1989. The four-dimensional nature of lotic ecosystems. *Journal of the North American Benthological Society* 8:2–8.

Wilby, A., and M.B. Thomas. 2002. Natural enemy diversity and pest control: Patterns of pest emergence with agricultural intensification. *Ecology Letters* 5:353–360.

9

Cascading Effects of Deforestation on Ecosystem Services Across Soils and Freshwater and Marine Sediments

Phillip Ineson, Lisa A. Levin, Ronald T. Kneib, Robert O. Hall, Jr., Jan Marcin Weslawski, Richard D. Bardgett, David A. Wardle, Diana H. Wall, Wim H. van der Putten, and Holley Zadeh

The full assessment of the impacts of ecosystem management and disturbances on the provision of ecosystem services would be a comparatively simple process if ecosystems were totally self-contained and independent. Simple monitoring, observation, and assessment within a system would inform us of the implications of our management activities, with the benefits and costs being simply quantified and traded. In that simple case, increases in yield could be weighed against biodiversity loss, the ability of an ecosystem to buffer against flooding could be traded for loss of its amenity value, and so on. However, this simplistic approach is clearly unrealistic, because "secondary" effects to other systems are an almost inevitable consequence of any adopted management policy or disturbance.

One simple and seemingly objective means of assessing the full economic and cultural value of ecosystems to humans is to consider the provision of ecosystem services (Daily et al. 2000). The ecosystem services of terrestrial, freshwater, and marine soils and sediments have been developed thematically in previous chapters of this volume. With our increasing appreciation of all the previously unquantified benefits that ecosystems have for humans, there is also an increasing requirement for techniques to formalize the impacts of human activities on these services. As outlined in previous SCOPE reports (Wall Freckman et al. 1997; Wall et al. 2001a; Wall et al. 2001b) there is a need to be able to extend this approach to interconnections between ecosystems, to quantify how

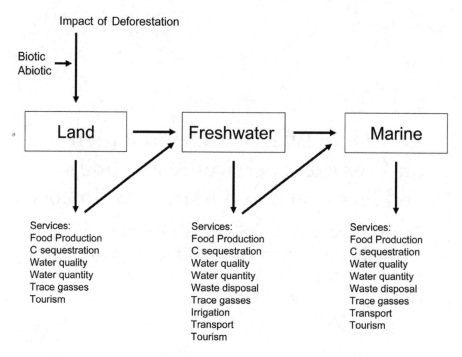

Figure 9.1. Overview of the potential impacts of deforestation on ecosystem services provided by cascading domains, showing the linkages between domains

altering the management of, or inflicting a major disturbance on, one system can have "downstream" effects on the provision of ecosystem services in other interconnected, but spatially separated, ecosystems and domains.

We have chosen to focus here on deforestation as an example of one such potentially interconnected series of cross domain impacts, because this issue has already been investigated from the perspective of each of the three domains (terrestrial, freshwater, marine) but these have never been formally linked together. Figure 4.1 in Chapter 4 shows the overall nature of these cascading links as a result of deforestation, while Figure 9.1 summarizes the approach used in the current evaluation; the objective is to strengthen our understanding of the importance of the links between the domains rather than simply emphasize the within-domain impacts.

However, such an analysis could be almost infinite if one considered all types of forests and all implications for terrestrial and aquatic species, their interactions, biodiversity, trophic food chains, and so on. Therefore, we have limited the scope of the discussion to deforestation of temperate systems and to those aspects directly linked to provision of ecosystem goods and services (Figure 9.1). As ecologists specifically interested in the biodiversity of soils and sediments, we have also focused on the role of biodiversity in ecosystem function, considering how changes in these biological parameters directly affect humans through the provision of ecosystem services. In providing this specific example, we utilize a structured and logical template for assessing the cascading effects of management on one ecosystem through other, spatially separated, domains.

Forests and Forest Soils: Deforestation Impacts

Despite the very diverse nature of forest soils, some gross generalizations about the structure and biological activity can be traced back to the original distinction between mull and mor made by Muller in 1887. The careful examination of soil humus forms, initially advocated by Muller (1887) and further developed by such researchers as Kubiena (1953) and Topoliantz et al. (2000), has reconfirmed the crucial interaction between soil parent material, faunal activity, and plant litter quality in determining the development of soil organic matter and in the value of the concepts of mull and mor humus.

Soils with mull humus typically form on calcium-rich, moderately or well-drained clayey parent material, generally beneath grassland and deciduous forest. The mull humus form is often mildly acidic, due to leaching of base cations (e.g., calcium) down the soil profile, and is characterized by intimate mixing of the surface organic and upper mineral A horizon. This mixing results from a high abundance and activity of soil biota, especially earthworms, and also leads to enhanced rates of decomposition, nutrient availability, and plant growth.

In contrast, in coniferous forests and other systems that produce poorer litter quality, there is an obvious build-up of organic matter at the soil surface: the so-called mor humus. These mor humus forms are typically acidic in nature and are characterized by low rates of decomposition and plant nutrient availability, differing both biologically and chemically from the mull form. Biologically, it is typical for the microbial biomass of these mor soils to be strongly dominated by fungi (rather than bacteria), and the fauna are characterized by high numbers of microarthropods (mites and Collembola), but with an absence of earthworms. Chemically, these humus forms are quite different: mull has a lower fulvic acid fraction, with a higher concentration of calcium and higher base exchange capacity in the humic fraction. Mor soils, by contrast, are characterized by less completely degraded organic matter, with an abundance of lignin and hemicellulose, resulting in a more fibrous texture. Both base cation content and exchange capacities tend to be much lower in mor than in mull soils.

Although providing a generalized starting point for describing humus forms in forest soils, it must be appreciated that many inter-grades and variations from these basic forms exist. For example, Kubiena (1953) extended these divisions by describing a series of important common transition forms, particularly that of "moder" humus, which superficially resembles a mor in having a sharp distinction between organic and

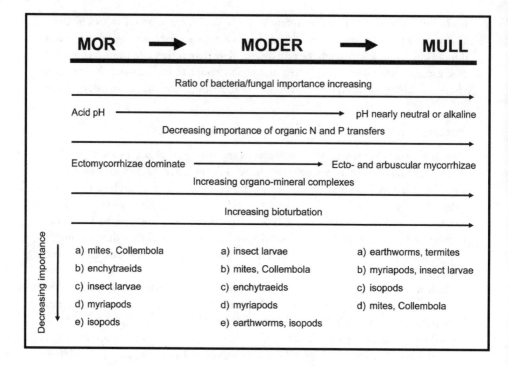

Figure 9.2. The relationship between humus types and fauna in forest ecosystems (adapted from Wallwork 1970).

mineral horizons, but contains large numbers of insect larvae and some restricted species of acid-tolerant earthworms. Thus, the differing humus forms in forest soils are a reflection of biological activity. Some of these differences are summarized in Figure 9.2, which has been modified from Wallwork (1970).

Forests have a number of key and unique features that determine the development of soils and associated biota. First, forests are among the most productive terrestrial ecosystems in the world, with both high rates of annual net primary productivity and the greatest standing biomasses; global terrestrial net primary production is estimated at around 60 PgC a^{-1}, of which 30 percent is accounted for by forests (see IPCC 2001; Saugier et al. 2001; Gower 2002). The input of organic matter into the soil occurs both as root input and aboveground litter fall, and the fate of these different litter materials is influenced both by their physical placement into the soil and by their intrinsic differences in quality. In particular, the input of woody remains to the forest floor is important both as a substrate and habitat for decomposers, and it has been argued that the input of plant material from "k strategy" plants may be reflected in the "r or k strategy" of the soil population (Heal & Ineson 1984). It has been also argued that an increase in the abundance of litters rich in lignin and more recalcitrant molecules requires a more complex decomposer community.

Forest soils provide an important habitat for a wide range of soil fauna, but forest carbon turnover is actually dominated by microbial activity, with carbon accumulation on the forest floor being the resultant between net primary productivity (above- and below-ground) and decomposition. The microbial community of the upper soil layers in forests is dominated by fungal-based rather than by bacterial-based food webs, which are highly capable of degrading the more recalcitrant components of the incoming litter and are less sensitive to the acidic conditions that frequently develop in the organic rich upper soil horizons (see Chapters 2 and 5, this volume). Fungi tend to be more resistant to drought, and they are well adapted to the changing microclimate in the litter layer through the production of both asexual and sexual sporing structures. However, the sensitivity of mycorrhizal fungi colonization to summer drought in grassland soils has recently been convincingly demonstrated by Staddon et al. (2003). Basidiocarps (reproductive bodies) produced by both saprotrophic and mycorrhizal fungi are an important direct ecosystem product, being a prized food resource across a wide range of human cultures.

Abiotically, one of the most important effects that deforestation and land-use change have on water draining from catchments is in sediment losses. Particulate loading into streams can be conveniently split into two divisions: suspended load and bedload. *Suspended load* is the sediment in transport, which is buoyed up by the movement and flow of the water and includes particles from clay through to fine sand. Typically, this is the material that can be collected in a "bucket" or "gulp" sampler, which can have a marked effect on the turbidity of the water. In contrast, the *bedload* is the particulate matter that rolls down the stream while remaining in contact with the channel bed, and which is actually quite difficult to quantify.

Bonin et al. (2003) collected fine benthic organic matter from a number of catchments in the United States and concluded that previous timber harvesting had a major impact on the sediment load and sediment biological activity at the base of the catchments. Indeed, these researchers found that the resulting fine organic material had higher mineralizable and extractable nitrogen when derived from the less-disturbed sites, and microbial activity (as represented by a wide assortment of assay techniques) in the harvested catchments tended to be greater. More dramatically, the historic felling of forests in New Zealand has been linked to major landslides that, in turn, have contributed significantly to the development of sediment in depositional basins both within the freshwater and nearshore and offshore zones of coastal systems (see Glade 2003, and below). A less obvious but more insidious problem resulting from deforestation is associated with the construction and materials associated with road building. When considered in terms of changes to the management of a forest, road-building practices have the potential to result in major changes to soils and to the freshwater ecosystem (Motha et al. 2003) at many temporal and spatial scales (see also Chapter 8, Lavelle et al.).

The provision by forests of ecosystem goods and services has been discussed throughout this volume, and a brief summary of the impacts of deforestation within the terrestrial domain is provided below. It should be noted, however, that these impacts are necessarily generalized, since specific site factors such as the scale, location, residue, and post-disturbance vegetation regime may be critical in controlling impacts. One extreme illustration of this is provided in Figure 9.3, which emphasizes the point that the spatial distribution of tributaries, even with an equivalent catchment area, can impact a receiving marine domain in very different and idiosyncratic ways, depending on details such as branching structure and number of lakes within the river system.

1. *Food provision.* Although at one time terrestrial forests were a major source of human nutrition, the development of agriculture has meant that forests no longer occupy such a central role. However, as a habitat for birds and mammals, forests currently represent an important source of meat for those human cultures still dependent on hunting, while even societies that rely heavily on agriculture still value the edible fungi and fruits growing wild in forests. Forest felling can have major effects on forests as sources of foods, increasing access for hunters of larger mammals and increasing the provision of forest margins and glades suitable for wild boar (Brownlow 1994). With respect to game birds, the traditional view is that the creation of forest edges increases the number of game birds, but this has been challenged by Temple and Flaspohler (1998). Thus, many of the effects of deforestation on human nutrition are positive, since the provision of appropriate forest clearings, edges, and rides can encourage, for example, species of mammals and fungi not found in intact forests, and may actually improve the habitat for specific mammals. Indeed, Termorshuizen and Schaffers (1991) suggested that clear-cutting was a significant factor in increasing fungal abundance due to the provision of disturbed, less polluted, soil layers.

2. *Carbon sequestration.* Soil biota are a crucial determinant of carbon sequestration and mixing in forest soils, with greater carbon accumulation in mor than mull systems. However, carbon stored in mor humus may be more vulnerable to deforestation than carbon held in mull humus, since the stabilization of organic matter onto minerals is more pronounced in mull soils. Earthworm activity is specifically associated with the intimate mixing of soil organic matter and minerals, which has major consequences for the sequestration of carbon. A large component of carbon storage in forest floors is maintained in trunks and woody debris, and the impacts of deforestation on carbon storage depend on how much of this material is left after felling. However, the net effect of deforestation sometimes results in soil carbon loss, largely because canopy removal results in a greater water and heat flux to the soil with increased decomposition together with a reduction of litter inputs after the initial felling.

3. *Water quantity.* Reduced rooting, together with an associated drop in mycorrhizal fungal biomass and less canopy interception, leads to an increase in soil moisture content and greater runoff. Water yields can be significantly greater in streams draining felled catchments (see below).

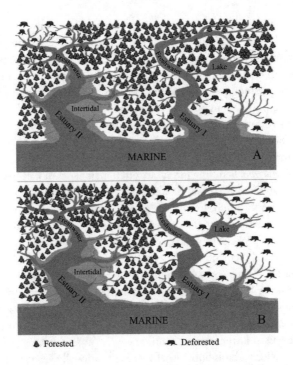

Figure 9.3. Diagrammatic representation showing how the physical location and scale of deforestation is important in controlling impacts on freshwater and marine sediments. The top panel represents a forest system with selective cutting, and the bottom panel represents a forest with clear-cutting. The level of intensity of the disturbance can impact freshwater and marine sediments, and the cascading effects of sedimentation, quite differently. For in-depth discussion of spatial and temporal scale, see Chapter 8.

4. *Water quality.* Deforestation exerts a major influence on decomposition processes, largely through the changes in leaf fall litter, heat, and water flux to the soil. The removal of tree root activity, together with the large underground litter input from roots, reduces water lost through evapotranspiration and also provides considerable nutrient and detrital input. In a review of the impacts of forest management practices on forest carbon stores, Johnson (1992) concluded that there were no consistent general trends of carbon change after harvesting, although lower soil carbon stocks occurred if the harvesting was followed by intense burning or cultivation. Although a few studies have shown large net losses, most studies show no significant change. Yanai et al. (2003) has highlighted the technical difficulties associated with detecting forest carbon store changes. Increases in dissolved organic carbon (DOC) and sediment load, together with increased concentrations of nitrate (see Binkley 2001), and cations have been frequently observed and experimentally verified. There is generally a transient increase in soil ammonium concentrations after

felling, but this often fails to manifest as a change in stream waters. A central feature here is the process of nitrification, which not only serves to produce the highly mobile anion NO_3^- but may also result in acidification under inappropriate management (Neal et al. 1998). The organisms associated with nitrification in acid soils are poorly understood, yet modern molecular techniques are demonstrating that the characteristic populations of these autotrophic bacteria differ between domains and habitats within domains. Increased stream water concentrations of Al^{3+} are also a consequence of felling on acidic soils, potentially having a deleterious impact on both soil and aquatic biota.

5. *Trace gas production.* Standing forests are often major sites of CH_4 oxidation within a landscape, and forest removal is often linked to a reduction in the capacity of soils to remove this important greenhouse gas. The mechanisms behind these changes are poorly understood, but are associated with the soil bacteria specifically known to carry out the process. In contrast, soil waterlogging associated with felling may result in increased methanogenesis, associated with the anaerobic decomposition of organic matter. Wetter soils, with a greater NO_3 concentration and more available carbon from the felling residue, all interact to increase denitrification rates, with N_2O being a common product under the acidic conditions found in forest soils. A significant quantity of dissolved N_2O may end up in the streams, being subsequently de-gassed farther downstream.

6. *Recreation.* Felled forest is both visually and recreationally less valuable than standing forest, with hikers, horse riders, and day-trippers preferring intact forest stands with good road and ride access. Other users, such as hunters, have slightly different preferences and prefer some open space within the forest. However, overall, recreation is reduced in recently felled forest. A summary of changes in ecosystem services and consequences for the outputs moving into linked freshwater systems is provided in Table 9.1, section A.

Freshwater Ecosystems: Deforestation Impacts

Impacts of deforestation on stream assemblages and ecosystem processes are well known, but they have not been considered explicitly within the context of the ecosystem services provided by stream benthos. There is a wealth of observations and experimental data on the impacts of deforestation on the physical, chemical, and biological characteristics of streams, potentially leading to changes in provision of ecosystem goods and services by the freshwater system. However, it must be recognized that not all deforestation is the same, and the impacts are closely linked to the actual management practices used in the felling process. The extent of the area and actual location of deforestation, whether principally in the upper or lower reaches of the catchment, also have a major influence on potential freshwater impacts (see Figure 9.3).

Streams receive increased sediment loads following tree removal (often from road building), which alters geomorphology because depositional areas contain a higher proportion of fine sediments (Platts et al. 1989; Waters 1995). The chemical composition of groundwater flowing from soils to streams can change radically with increased solutes such as nitrate, base cations, and lower pH (Likens et al. 1970). Water temperatures may increase because of the loss of overhanging vegetation (Johnson & Jones 2000) and the considerable reduction of leaf litter (Webster et al. 1990) during the short period before regrowth. Wood inputs will decrease (Bragg 2000), which alter stream geomorphology, since wood may play an essential role in determining riffle-pool structure and sediment storage in streams (e.g., Montgomery et al. 1996).

Stream biota can play a strong role in providing such ecosystem services as juvenile fish production, nutrient transformation, and recreation (see Covich et al., Chapter 3; Wall et al. 2001). Altering these biota via forest removal may affect some of the ecosystem services that streams provide, with reduced litter and wood inputs combined with increased sediment and solute loads, resulting in changes in stream biota. Invertebrate assemblage structure shifts from large shredder-dominated assemblages to increased scrapers (Gurtz & Wallace 1984), though there can be higher secondary production of macroinvertebrates well after the disturbance (Stone & Wallace 1998). Responses to logging are habitat specific; depositional areas may have lower abundance of invertebrates following logging, presumably from increased sediment, while bedrock reaches may experience higher invertebrate density due to increased light (Gurtz & Wallace 1984).

Responses of fish populations and production to deforestation are complex. Increased algal production from higher light input may increase fish production, largely as a result of higher invertebrate production (Bilby & Bisson 1992). However, long-term loss of wood inputs will reduce habitat diversity and fish biomass in streams, despite the short-term increases in light and algal productivity. For example, streams in British Columbia had four to ten times higher salmonid biomass when woody debris was abundant (Fausch & Northcote 1992). Sediment can fill spawning gravel, lowering salmonid recruitment success (Waters 1995).

Deforestation can alter ecosystem services provided by the stream benthos. If the stream has migratory salmonids, there may be strong negative effects on recruitment of smolts (Table 9.1, section B) and much of the decrease of salmon populations in coastal streams has been attributed to sediment impacts from deforestation (review in Waters 1995). Thus, removal of forest production may dramatically affect the stream's ability to recruit salmon, which will lower overall productivity, and, in turn, harm recreational fisheries.

Deforestation may also decrease the stream's capacity to remove nutrients from the water column, resulting from both abiotic and biotic changes. Increased nutrient concentrations may saturate biotic uptake (Dodds et al. 2002), which may proportionally increase loads of dissolved inorganic nutrients. Loss of coarse woody debris may also lower nutrient uptake and storage (Bilby & Likens 1980; Valett et al. 2002), and

Table 9.1. Summary of consequences of deforestation on ecosystem services in terrestrial, freshwater, and marine domains with emphasis on soil and sediment biodiversity.

A rank from −3 (strong disservice) through 0 (neutral) to +3 (strong service) is given for each good or service, indicating its value to human societies.

Ecosystem Service	*A: Terrestrial*			*Imports to Fresh Water*	*B: Freshwater*			*Imports to Ocean*	*C: Marine*		
	Rank	Biotic	Abiotic		Rank	Biotic	Abiotic		Rank	Biotic	Abiotic
Food production	−1	mushrooms	pH		−3	deteriorated spawning grounds	sediments cover gravel		?	increased sediment builds nursery grounds	salinity increased
		hunting	soil texture			whirling disease (tubificids)	increased temperature			declines in filter feeding bivalves	higher turbidity
		nuts	moisture			shredders decline				increased primary production	more nutrients
		honey	soil fertility			loss of insect taxa				increases in deposit feeders	sediment structure
		berries	light			algae/deposit feeders increased				marine seep	
Carbon sequestration	−3	increased decomposition	increased temperature		0				1	more marsh vegetation	DOC locked up

							more algae	clays
Water quality	−3							
other fauna increase (see narrative)								
nitrifying bacteria	increased moisture	loss of structure	−3	algae remove nitrate	more DOM	1	more particle trapping by plants	increased turbidity
decomposers	sediment mobilization	sediment mobilization		denitrification	increased sediment	sediment mobilization		increased nitrogen
reduced roots	increased NO_3	increased NO_3		reduced filterers	more light	increased NO_3		changing salinity gradient
reduced fungal hyphae	decreased SO_4^{--}	decreased SO_4^{--}		food web effects on microbes	increased temperature	more DOC		
loss of leaf litter	increased DOC, DOP, DOM	increased DOC, DOP, DOM		change in C export	more water			
	base cations increased	base cations increased						
	pH decreased	pH decreased						
	increased N_2O	increased N_2O						
	increased temperature	increased temperature						
	increased light/UV	increased light/UV						
	aluminum increased	aluminum increased						

(continued)

Table 9.1. (continued)

Ecosystem Service	A: Terrestrial			Imports to Fresh Water	B: Freshwater			Imports to Ocean	C: Marine		
	Rank	Biotic	Abiotic		Rank	Biotic	Abiotic		Rank	Biotic	Abiotic
Water quantity	2	decreased root and fungal hyphae	reduced water retention higher discharge peak		0		abiotic routing				
Trace gas production	−2	increased N_2O reduced methane oxidation	increased moisture increased temperature		−1	N_2O production	physical degassing of N_2O	physical degassing of N_2O	−1	methanogenesis	
Recreation	−2	tree loss fewer hikers			−1	less fishing aesthetic value declines					
Biological control									2	disease control limits invasive species migration up estuary	increased freshwater decreased salinity

increased fine sediment may fill interstices within stream substratum, decreasing hyporheic exchange (Morrice et al. 1997); this may reduce key biotic processes such as denitrification. However, increased fine inorganic sediment may increase uptake and storage of ions such as phosphate, which will bind to increased fine sediment (Meyer 1979). Increased sediment may reduce numbers of filter-feeding insects, which may lower downstream transport of particle carbon. In some cases, however, the reduction of filter feeders would not have as serious effects on transport of particle carbon as would the reduction of estuarine bivalves (Wallace & Merritt 1980; Caraco et al. 1997; Jassby et al. 2002; but see also Crowl et al. 2001).

The net effect of rivers and associated riparian zones is to lower the dissolved nutrient concentrations that are exported to estuaries (Alexander et al. 2000), with deforestation having to occur over a large portion of the watershed in order to substantially reverse this process. Given that nutrient concentrations decrease following regrowth of vegetation, and that large watersheds are cut sequentially, it is unlikely that increases in nutrients would rival agricultural or atmospheric inputs. Similarly, much of the sediment transported to streams may be stored in stream channels, thereby attenuating the delivery of sediment to downstream ecosystems. It is possible that temporary losses of organic matter from allochthonous (non-local) inputs could lower productivity in downstream ecosystems if there is substantial transport of this material (e.g., Vannote et al. 1980). It is more likely that loss of instream retention structures (e.g., woody debris) will decrease the ability of the stream to retain organic matter, with streams from logged catchments potentially having higher rates of organic matter turnover and higher export than non-logged catchments, at the expense of benthic storage (Webster et al. 1990). Despite small-scale catchment studies of organic matter export, there are few tests of large-scale impacts of land use on organic imports in downstream ecosystems. The provision of ecosystem goods and services within the freshwater domain have been discussed throughout this volume, and a brief summary of the impacts of deforestation within the terrestrial domain is provided below.

1. *Food provision.* The quantity and quality of fresh waters have particularly marked effects on salmon production, with decreased salmon spawning due primarily to increased sediment input to streams associated with felling. Whirling disease may exacerbate problems. Other biological changes are exerted via the shredders, which decline with the increased algal blooms resulting from water-quality changes. Water-quality changes will increase the numbers of deposit feeders, and abiotic factors, such as increasing temperatures, may also be affected.

2. *Carbon sequestration.* Streams are not considered to be primary sites for C sequestration because of their relatively small area and low speed of flow. However, after felling it is anticipated that the net balance between increased runoff and increased sediment load will lead to a net increase in river sediment.

3. *Water quantity.* For the reasons presented in the terrestrial domain section (above),

the quantity of stream water will increase post-felling. This increased yield may be significant, and it is an important consideration when forested catchments are being managed largely for water provision.

4. *Water quality.* Most countries have recommended guidelines or legislation for forest-felling operations in areas where water from a catchment is important for human consumption. Specific limitations on stream-edge felling, road construction, pesticide usage, and so on are designed to protect water quality, with concentrations of NO_3, Al, dissolved organic matter, and sediment being critical indexes of water quality. The impairment of this important ecosystem service can impact different sectors depending on the chemical changes that occur. For example, high concentrations of nitrate are considered a risk to human health, while aluminum may be toxic to fish and other stream biota; high DOC and associated dark water color may give rise to additional processing or mixing costs.

5. *Trace gas production.* During transfer in the stream, some nitrogen will be removed from the water course by algae and also by the process of denitrification, resulting in the production of N_2O. Trace gases, including N_2O and CO_2 previously produced in the soil and dissolved in steam water, will be de-gassed from the stream into the atmosphere.

6. *Recreation and transport.* Deforestation is frequently associated with new roads, which may increase recreational access for fishing, canoeing, and so on. Major rivers are actually used to transport timber from logging sites. However, increased sediment loads and Al^{3+} may cause reductions in specific fish stock.

An overview of changes of ecosystem services and consequences for outputs toward freshwater systems is provided in Table 9.1 section B.

Marine Ecosystems: Deforestation Impacts

Key inputs into the coastal zone (estuaries and shelf) from upstream will include sediments, nutrients, pollutants, and, to a lesser extent, fresh water. They will have both positive and negative effects on the services provided by sediment habitats (salt marsh, mangrove, seagrass beds, and tidal flats) and by sediment-dwelling marine benthos. These services are discussed in more detail in Chapters 4 and 7, and are also reviewed by Snelgrove et al. (1997, 2000) and Levin et al. (2001). Briefly, they include ecological functions such as decomposition and nutrient regeneration; regulation of water fluxes, particles, and organisms; and habitat and nursery provision. They also provide more direct services: shoreline stabilization; water purification; the production of harvestable fish, shellfish, and plants; and trophic support for these fisheries. Key functional groups involved in providing these services include bacteria and fungi, shredders, suspension feeders, and bioturbators (see Wall et al. 2001b). Below we review a range of scenarios regarding altered sediment function in the coastal zone that might result from increased inputs of nutrients in sediments of freshwater associated with upland deforestation.

The life histories and health of many estuarine organisms are finely tuned to the quantity and timing of freshwater input to estuaries, and to the resulting salinity variations. Attrill (2002) has demonstrated that estuarine diversity in the Thames River Estuary is lowest where the variation is greatest (often in the zone where salinity averages 5–15 ppt). Drought usually increases estuarine salinity and may reduce habitability for upper estuary oligohaline organisms. Invasions by marine species and increased incidence of disease have been attributed to reduced freshwater inputs associated with drought, as the timing of reproduction or growth is finely tuned to winter or spring flooding. However, increases in freshwater input associated with inland deforestation are likely to be minor and these effects will be minimal in comparison with those of sediment and nutrient input.

Increased nutrient inputs to the estuary from freshwater sources (rivers, streams, and groundwater) may impair or enhance key ecosystem services, depending on the level of nutrient loading. Because nitrogen availability often limits primary production by phytoplankton, algae, and rooted vegetation, low-level increases in nitrate concentration could increase primary production. The sediment biota may respond with increased abundance of all functional groups, leading to greater trophic support for, and production of, fish and shellfish. Increased nutrient availability may yield enhanced growth and peat deposition by vascular plants (salt marsh plants, mangroves, and sea grasses). This would increase C sequestration, stabilize sediments, and improve water quality through slowed flows and particle removal. The spread of wetlands may also lead to elevated organic matter accumulation, greater microbial activity, and greater production of methane, a potent greenhouse gas. However, concurrent increase in turbidity may negate the positive effects of nutrients on primary production. High nutrient loading, in contrast, will lead to levels of phytoplankton production that quickly overload the system. Decay of unconsumed primary production will increase biological oxygen demand and generate hypoxic waters, a process termed *eutrophication*. A sequence of changes in the sediment biota may be expected, with large epibiota (e.g., crabs, shrimp, bottom fish) and subsurface dwelling forms (bivalves, echiurans, annelids) declining first, with loss of bioturbation and nutrient regeneration capacity. Small, opportunistic taxa that tolerate hypoxia construct tubes or shallow burrows in dense assemblages near the sediment-water interface (Pearson & Rosenberg 1978), and they have limited influence on the cycling or burial of organic matter (Rhoads & Boyer 1982). Diversity of species and functional groups will decline with increasing hypoxia. With severe eutrophication, the sea floor may be carpeted with dead plankton, fish, and other organisms. These will become covered by sulfur bacteria, and underlying sediments will be anoxic and azoic. Such conditions have been noted in the Baltic and North Sea in the past, and occur in summer in the coastal Gulf of Mexico (Rabalais et al. 2001). It is unlikely, however, that the release of nutrients and organic debris by deforestation alone would be of a magnitude sufficient to induce such severe eutrophication in coastal estuaries. Typically, nitrogen fertilizer and phosphate runoff from domestic and industrial use are contributing factors.

Steep rivers that cut through mountainous regions may deposit large amounts of terrestrial debris and fine organic matter on narrow continental shelves and slopes during flood events (Leithold & Hope 1999; Wheatcroft 2000). On the Eel River margin of northern California, deposits of woody debris coincide with the occurrence of methane seepage. On the slope at 500 m, distinctive infaunal assemblages are formed (Levin et al. 2003) that have carbon isotopic signatures consistent with a carbon source derived from terrestrial organic matter (Levin & Michener 2002). It is possible that deforestation effects on organic inputs could be felt on the shelf and slope where steep mountainous terrain abuts the coastal zone.

Suspended Sediment

Total suspended particulate matter (TSM) contains both organic particulate organic matter (POM) and inorganic particles transported with riverine systems to estuaries and the sea. Its content varies seasonally and geographically, ranging from 10 to 1000 mg TSM per cubic decimeter. As they reach the marine zone, suspended sediments undergo rapid transformation and are largely removed from the water column in the process of physicochemical flocculation, aggregation, and finally sedimentation. In open river mouths discharging the freshwaters directly to the full saline sea, as much as 95 percent of POM is sedimented in the immediate vicinity of river mouths, and only a small fraction of the original suspended load is transported to the shelf (Lisitzin 1999). Sedimentation rates in river mouths are extremely variable, but may range over 1500 g/m^2/day; similar values have been reported from vegetated salt wetlands and tidal flats. TSM loads may enter the estuaries both as bedload transported and surface-layer transported particles, depending on the hydrology, season, speed of the flow, and so on. In the first instance, the surface estuarine waters are relatively transparent (euphotic) and local primary production may be high. In other cases, the water transparency drops and local primary production is completely inhibited. The increase of suspended load and consequent increase of sedimentation have profound effects on the estuarine sedimentary biota.

The overall reaction of benthic assemblages follows the pattern described by Pearson and Rosenberg (1978) in disturbed or enriched marine sediments. Large, sessile, structure-forming species disappear and are replaced by small, mobile, surface-dwelling taxa. The increase of TSM load influences the services provided by estuarine-shelf biota. Effects include a drop in biomass of harvestable species or enhanced carbon sequestration by bacterial development on mass aggregates of organic matter. Other services, such as recreation and aesthetic values, are likely to be affected negatively due to the turbid water, decrease of fish production, and organic matter deposits.

The effects of terrestrial sediment deposition on New Zealand estuaries, as mentioned earlier, have been particularly well studied and are considered here in more detail. Extensive deforestation and other development-related activities have led to sediment inputs via upland landslides, sediment runoff, and river-carried material in New Zealand. Historical sediment deposition records indicate that New Zealand experienced massively increased sediment input to the rivers and estuaries with the arrival of humans on the island, and again with the arrival of Europeans, who were responsible for much of the deforestation (Hume & McGlone 1986). The most deadly sediment inputs occur in the form of clay-dominated subsoils, and flooding and landslides may result in layers of this material millimeters to centimeters thick being deposited into estuaries; single deposition events of up to 10 cm have been recorded (Thrush 2004). The clays are associated with low pH and often with refractory organic matter. Few animals are able to burrow through clay layers, and sediments beneath these layers become highly anoxic. Experimental deposition of clay-dominated sediments reveal significant negative impacts and slow rates of macrobenthic recovery (Norkko et al. 2002; Cummings et al. 2003; Hewitt et al. 2003; Thrush et al. in press). Even more chronic deposition events (1, 3, 5, 7 mm thick) can produce changes in the density of macrofaunal density of up to 50 percent (Thrush 2004). These effects include bivalve and polychaete mortality, as well as changes in feeding behavior, declines in growth density, and slowed recruitment, with effects lasting up to 18 months or more (Norkko et al. 2002). Only large burrowing crustaceans (crabs and shrimp) are able to cope with the high sediment loads. Microphytobenthos (benthic microalgae) that can migrate up through the layers of clay will survive, but those that cannot will die, diminishing the food supply for many grazers in sediments. Experiments reveal that species-level identification is necessary to evaluate consequences of sediment input for estuarine sediment communities. Within a single functional group (Spionidae, a family of surface-feeding polychaetes), sensitivity to clay deposition may differ greatly among species (Thrush et al. 2003, in press).

In New Zealand, sediment fecal bacteria often accompany sediment runoff from agricultural land. Together the sediments and bacteria have negative impacts on harvests of filter-feeding oysters (*Crassostrea gigas*) and mussels (*Perna canaliculus;* Hawkins et al. 1999) from marine farms as well as affecting recreational and traditional shellfish harvesting by the Maori population. Increased sedimentation and associated nutrient input may also account for expansion of mangroves in many of the harbors and estuaries in northern New Zealand (Dingwall 1984).

Below, in summary, we list some of the key impacts of deforestation across a range of marine and coastal ecosystems services:

1. *Food provision.* Water quality, including inorganic and organic solutes and, particularly in this case, sediment load from rivers, can have major effects on food provision by coastal and estuarine systems. These effects may be both positive and negative, ranging from potentially increasing the build-up of fish nursery grounds

through to declines in the number of filter-feeding bivalves. As outlined above, the sediment input via fresh water entering marine systems has particularly marked effects on salmon production.

2. *Carbon sequestration.* The continental margins play a major, but poorly quantified, role as burial sites of organic carbon derived from both oceanic and terrestrial sources. Perturbations in freshwater discharge alter the intensity and depth of penetration of thermohaline circulation, with consequences for the fate of carbon, the transfer of dissolved organic carbon and sediment, and the export of dissolved organic carbon and particulate organic matter.

3. *Water quantity.* For the reasons presented above, the quantity of stream water will increase post-felling.

An overview of changes of ecosystem services and consequences for outputs toward marine systems is provided in Table 9.1 section C.

Discussion and Conclusions

In this analysis we have necessarily restricted our considerations to the cross-domain impacts of one specific disturbance in one ecosystem type; however, some generic conclusions have emerged (Table 9.2). First, it is very clear that significant impacts can be readily passed on through a chain of domains, with important consequences for the downstream systems. For example, the logging of large forest areas can affect remote coastal marine domains, as seen in the clear example from New Zealand discussed above. The impacts on the sediment biota are central to the sensitivity of the system, with attendant consequences for ecosystem services of importance to humans. Again, we have restricted discussion to the felling of temperate forests but similar, and equally dramatic, effects have also been reported for nontemperate systems. For example, impacts of increased sediment loads on sensitive systems such as coral reefs (e.g., Dubinsky & Stambler 1996) are particularly noteworthy because of their large conservational, and increasingly recreational, value.

Second, it is also clear that a single type of disturbance may have differing impacts on the different receiving domains, and that these effects may not be simply proportionate to distance from the disturbance. For example, deforestation may lead locally to increased stream nitrate and aluminum concentrations with, respectively, consequences for local water potability and fish stocks. However, by the time these streams have merged with others and then entered the oceans, the effects may have become insignificant. The sensitivity of the organisms and processes in the receiving system, amounts of dilution, and existing load from other sources are all key in controlling the extent of the effects attributable to a specific land management change. It should be emphasized that the case for sediment load is very different from the case for solutes, with the coastal system effectively acting as an accumulation point for sedi-

Table 9.2. Net effects of deforestation on soil, freshwater, and marine domains.

Terrestrial	Freshwater	Marine
Loss of sediment	High sediment load	More wetland
Loss of nutrients	Increased NO_3 + other nutrients	Sediment load
Change in soil biota	More water	Nitrogen stimulation of production
Loss of carbon	Increased light and temperature	Altered benthic production
Increased temperature	Switch from autotrophy to heterotrophy	
	Alteration of sediment biota	

ment, while also containing organisms and processes sensitive to changes in these inputs. In the case of nitrate loading, nutrients in the coastal zone are normally dominated by fertilizer and effluent inputs, which largely dwarf any solute contributions arising from deforestation.

In their evaluation of the spatial appropriation of both marine and terrestrial ecosystem resources for the Baltic Sea drainage basin, Jansson et al. (1999) emphasized the critical interdependence between freshwater flows and the capacity of ecosystems to generate services, emphasizing the need for a holistic approach to watershed landscape management to avoid any unintentional effects and loss of services. We would argue that improved anticipation of unintentional effects across domains can be achieved using the framework described here, with a systematic assessment of likely cross-domain effects. Management options that lead to the lowest impairment of ecosystem services, balanced across all domains, should be selected.

The decision to limit the current evaluation to deforestation and temperate systems, while making the process tractable, has also limited the extent to which any resulting conclusions can be generalized. However, a brief consideration of studies in nontemperate regions actually lends support to the general conclusions. For example, the role of deforestation in increasing sediment loads and, more specifically, mercury contamination in lacustrine sediments in the central Amazon has been reported by Roulet et al. (2000), indicating a major impact of anthropogenic disturbance on mineral and organic cycling across the entire region. Studies examining the causes for the substantial quantitative and qualitative changes in deposited sedimentary organic matter in major tributaries of the Amazon also strongly implicate increased deforestation as a major cause (Farella et al. 2001), although the consequences are less fully described. As mentioned above, increased sediment load from deforestation and topsoil erosion have been linked to substantial marine changes in specific nontemperate marine systems, including coral death and eutrophication (Dubinsky & Stambler 1996). Downing et al.

(1999) argued that alterations in the N cycle, resulting from forest disturbance in tropical systems, will have much greater impacts on tropical aquatic ecosystems than on temperate equivalents, largely because of the greater importance of nutrient limitations in
these systems.

Thus, it appears to be the case that the general conclusions arising from the specific
case study will extend to other biome types; however, we have not addressed the question of the extent to which the concepts developed here can be applied to other environmental changes. The evidence from the current evaluation would suggest that
changes resulting from perturbing a single domain will frequently be seen as impacts in
other domains, and that different domains may respond surprisingly different to the
same environmental change—it is clear that one domain's meat may be another
domain's poison.

Literature Cited

Alexander, R.B., R.A. Smith, and G.E. Schwarz. 2000. Effect of stream channel size on
the delivery of nitrogen to the Gulf of Mexico. *Nature* 403:758–761.

Attrill, M.J. 2002. A testable linear model for diversity trends in estuaries. *Journal of Animal Ecology* 71:262–269.

Bilby, R.E., and P.A. Bisson. 1992. Allochthonous vs. autochthonous organic matter contributions to the trophic support of fish populations in clear-cut and old-growth
forested streams. *Canadian Journal of Fisheries and Aquatic Sciences* 49:540–551.

Bilby, R.E., and G.E. Likens. 1980. Importance of organic debris dams in the structure
and function of stream ecosystems. *Ecology* 61:1107–1113.

Binkley, D. 2001. *Patterns and processes of variation in nitrogen and phosphorus concentrations in forested streams.* NCASI Technical Bulletin #836, Research Triangle Park, NC.

Bonin, H.L., R.P. Griffiths, and B.A. Caldwell. 2003. Nutrient and microbiological characteristics of fine benthic organic matter in sediment settling ponds. *Freshwater Biology*
48:1117–1126.

Bragg, D.C. 2000. Simulating catastrophic and individualistic large woody debris recruitment for a small riparian system. *Ecology* 81:1383–1394.

Brownlow, M.J.C. 1994. Towards a framework of understanding for the integration of
forestry with domestic pig (*Sus scrofa domistica*) and European wild boar (*Sus scrofa
scrofa*) husbandry in the United Kingdom. *Forestry* 67:189–218.

Caraco, N.F., J.J. Cole, P.A. Raymond, D.L. Strayer, M.L. Pace, S.E.G. Findlay, and D.T.
Fischer. 1997. Zebra mussel invasion in a large, turbid river: Phytoplankton response to
increased grazing. *Ecology* 78:588–602.

Crowl, T.A., W.H. McDowell, A.P. Covich, and S.L. Johnson. 2001. Freshwater shrimp
effects on detrital processing and localized nutrient dynamics in a montane, tropical
rainforest stream. *Ecology* 82:775–783.

Cummings, V.J., S.F. Thrush, J.E. Hewitt, and S. Pickmere. 2003. Terrestrial sediment
deposits in soft-sediment marine areas: Sediment characteristics as indicators of habitat
suitability for recolonization. *Marine Ecology Progress Series* 253:39–54.

Daily, G.C., T. Soderqvist, S. Aniyar, K. Arrow, P. Dasgupta, P.R. Ehrlich, C. Folke, A.
Jansson, B.O. Jansson, N. Kautsky, S. Levin, J. Lubchenco, K.G. Maler, D. Simpson,

D. Starrett, D. Tilman, and B. Walker. 2000. Ecology: The value of nature and the nature of value. *Science* 289:395–396.

Dingwall, P.R. 1984. Overcoming problems in the management of New Zealand mangrove forests. In: *Physiology and Management of Mangroves,* edited by H.J. Teas, pp. 13:97–106. The Hague, Dr. W. Junk Publishers.

Dodds, W.K., A.J. Lopez, W.B. Bowden, S. Gregory, N.B. Grimm, S.K. Hamilton, A.E. Hershey, E. Marti, W.H. McDowell, J.L. Meyer, D. Morrall, P.J. Mulholland, B.J. Peterson, J.L. Tank, H.M. Valett, J.R. Webster, and W. Wollheim. 2002. N uptake as a function of concentration in streams. *Journal of the North American Benthological Society* 21:206–220.

Downing, J.A., M. McClain, R. Twilley, J.M. Melack, J. Elser, N.N. Rabalais, W.M. Lewis, R.E. Turner, J. Corredor, D. Soto, A. Yanez-Arancibia, J.A. Kopaska, and R.W. Howarth. 1999. The impact of accelerating land-use change on the N-cycle of tropical aquatic ecosystems: Current conditions and projected changes. *Biogeochemistry* 46:109–148.

Dubinsky, Z., and N. Stambler. 1996. Marine pollution and coral reefs. *Global Change Biology* 2:511–526.

Farella, N., M. Lucotte, P. Louchouarn, and M. Roulet. 2001. Deforestation modifying terrestrial organic transport in the Rio Tapajos, Brazilian Amazon. *Organic Geochemistry* 32:1443–1458.

Fausch, K.D., and T.G. Northcote. 1992. Large woody debris and salmonid habitat in a coastal British Columbia stream. *Canadian Journal of Fisheries and Aquatic Sciences* 49:682–693.

Glade, T. 2003. Landslide occurrence as a response to land use change: A review of evidence from New Zealand. *Catena* 51:297–314.

Gower, S.T. 2002. Productivity of terrestrial ecosystems. In: *Encyclopedia of Global Change, Vol. 2,* edited by H.A. Mooney and J. Canadell, pp. 516–521. Oxford, Blackwell Scientific.

Gutrz, M.E., and J.B. Wallace. 1984. Substrate mediate response of invertebrates to disturbance. *Ecology* 65:1556–1569.

Hawkins, A.J.S., M.R. James, R.W. Hickman, and S.M.W. Hatton. 1999. Modelling of suspension-feeding and growth in the green-lipped mussel *Perna canaliculus* exposed to natural and experimental variations in seston availability in the Marlborough Sounds, New Zealand. *Marine Ecology Program Series* 191:217–232.

Heal, O.W., and P. Ineson. 1984. Carbon and energy flow in terrestrial ecosystems: Relevance to microflora. In: *Current Perspectives in Microbial Ecology,* edited by M.J. Klug, and C.A. Reddy, pp. 394–404. Washington, DC, American Society for Microbiology.

Hewitt, J.E., V.J. Cummings, J.I. Ellis, G.A. Funnell, A. Norkko, T.S. Talley, and S.F. Thrush. 2003. The role of waves in the colonization of terrestrial sediments deposited in the marine environment. *Journal of Experimental Marine Biology and Ecology* 290:19–48.

Hume, T.M., and M.S. McGlone 1986. Sedimentation patterns and catchment use change recoded in the sediments of a shallow tidal creek, Lucas Creek, Upper Waitemata Harbour, New Zealand. *New Zealand Journal of Marine and Freshwater Research* 20:677–687.

IPCC. 2001. Intergovernmental Panel on Climate Change. *Climate Change 2001: The Scientific Basis.* Cambridge, UK, Cambridge University Press.

Jansson, A., C. Folke, J. Rockstrom, and L. Gordon. 1999. Linking freshwater flows and ecosystem services appropriated by people: The case of the Baltic Sea drainage basin. *Ecosystems* 2:351–366.

Jassby, A.D., J.E. Cloern, and B.E. Cole. 2002. Annual primary production: Patterns and mechanisms of change in a nutrient-rich tidal ecosystem. *Limnology and Oceanography* 47:698–712.

Johnson, D.W. 1992. Effects of forest management on soil carbon storage. *Water Air and Soil Pollution* 64:83–120.

Johnson, S.L., and J.A. Jones. 2000. Stream temperature responses to forest harvest and debris flows in western Cascades, Oregon. *Canadian Journal of Fisheries and Aquatic Sciences* 57:30–39.

Kubiena, W.L. 1953. *The Soils of Europe*. London, Thomas Murby and Co.

Leithold, E.L., and R.S. Hope. 1999. Deposition and modification of a flood layer on the northern California shelf: Lessons from and about the fate of terrestrial particulate organic carbon. *Marine Geotechnology* 154:183–195.

Levin, L.A., D.A. Boesch, A. Covich, C. Dahm, C. Erseus, K.C. Ewel, R.T. Kneib, A. Moldenke, M.A. Palmer, P. Snelgrove, D. Strayer, and J.M. Weslawski. 2001. The function of marine critical transition zones and the importance of sediment biodiversity. *Ecosystems* 4:430–451.

Levin, L.A., and R. Michener. 2002. Isotopic evidence of chemosynthesis-based nutrition of macrobenthos: The lightness of being at Pacific methane seeps. *Limnology and Oceanography* 47:1336–1345.

Levin, L.A., W. Ziebis, G.F. Mendoza, V.A. Growney, M.D. Tryon, K.M. Brown, C. Mahn, J.M. Gieskes, and A.E. Rathburn. 2003. Spatial heterogeneity of macrofauna at northern California methane seeps: The influence of sulfide concentration and fluid flow. *Marine Ecology Progress Series* 265:123–139.

Likens, G.E., F.H. Bormann, N.M. Johnson, D.W. Fischer, and R.S. Pierce. 1970. Effects of forest cutting and herbicide treatment on nutrient budgets in the Hubbard Brook watershed-ecosystem. *Ecological Monographs* 40:23–47.

Lisitzin, A.P. 1999. The continental-ocean boundary as a marginal filter in the world oceans. In: *Biogeochemical Cycling and Sediment Ecology. NATO ASI Series No. 2*, edited by J.S. Gray, W. Ambrose, Jr., and A. Szaniawska, pp. 69–103. Boston, Kluwer Academic Publishers.

Meyer, J.L. 1979. The role of sediments and bryophytes in phosphorus dynamics in a headwater stream ecosystem. *Limnology and Oceanography* 24:365–375.

Montgomery, D.R., T.B. Abbe, J.M. Buffington, N.P. Peterson., K.M. Schmidt, and J.D. Stock. 1996. Distribution of bedrock and alluvial channels in forested mountain drainage basins. *Nature* 381:587–589.

Morrice, J.A., H.M. Valett, C.N. Dahm, and N.E. Campana. 1997. Alluvial characteristics, groundwater-surface water exchange, and hydrologic retention in headwater streams. *Hydrological Processes* 11:253–267.

Motha, J.A., P.J. Wallbrink, P.B. Hairsine, and R.B. Grayson. 2003. Determining the sources of suspended sediment in a forested catchment in southeastern Australia. *Water Resources Research* 39:1056.

Muller, P.E. 1887. *Studien über die natürlichen Humusformen und deren Einwirkungen auf Vegetation und Boden*. New York, Springer-Verlag.

Neal, C., B. Reynolds, J.K. Adamson, P.A. Stevens, M. Neal, M. Harrow, and S. Hill.

1998. Analysis of the impacts of major anion variations on surface water acidity particularly with regard to conifer harvesting: Case studies from Wales and northern England. *Hydrology and Earth System Sciences* 2:303–322.

Norkko, A., S.F. Thrush, J.E. Hewitt, V.J. Cummings, J. Norkko, J.I. Ellis, G.A. Funnell, D. Schultz, and I. MacDonald. 2002. Smothering of estuarine sandflats by terrigenous clay: The role of wind-wave disturbance and bioturbation in site-dependent macrofaunal recovery. *Marine Ecology Progress Series* 234:23–41.

Pearson, T.H., and R. Rosenberg. 1978. Macrobenthic succession in relation to organic enrichment and pollution in the marine environment. *Oceanography and Marine Biology Annual Review* 16:229–311.

Platts, W.S., R.J. Torquemada, M.L. Henry, and C.K. Graham. 1989. Changes in the salmon spawning and rearing habitat from increased delivery of fine sediment to the South Fork Salmon River, Idaho. *Transactions of the American Fisheries Society* 118: 274–283.

Rabalais, N.N., D.E. Harper, and R.E. Turner. 2001. Responses of nekton and demersal and benthic fauna to decreasing oxygen concentrations. In: *Coastal Hypoxia,* edited by N. Rabalais and R.E. Turner, pp. 115–128. Washington, DC, American Geophysical Union.

Rhoads, D.C., and L.F. Boyer. 1982. The effects of marine benthos on physical properties of sediments. In: *Animal-Sediment Relations,* edited by T.L. McCall and M.Y. Tevesz, pp. 3–52. New York, Plenum.

Roulet, M., M. Lucotte, J.R.D. Guimaraes, and I. Rheault. 2000. Methylmercury in water, seston, and epiphyton of an Amazonian river and its floodplain, Tapajos River, Brazil. *Science of the Total Environment* 261:43–59.

Saugier, B., J. Roy, and H.A. Mooney. 2001. Estimations of global terrestrial productivity: Converging toward a single number? In: *Terrestrial Global Productivity,* edited by J. Roy, B. Saugier, and H.A. Mooney, pp. 543–557. San Diego, California, Academic Press.

Snelgrove, P.V.R., M.C. Austen, G. Boucher, C. Heip, R.A. Hutchings, G.M. King, I. Koike, P.J.D. Lambshead, and C.R. Smith. 2000. Linking biodiversity above and below the marine sediment-water interface. *BioScience* 50:1076–1088.

Snelgrove, P., T.H. Blackburn, P.A. Hutchings, D.M. Alongi, J.F. Grassle, H. Hummel, G. King, I. Koike, P.J.D. Lambshead, N.B. Ramsing, V. Solis-Weiss, and D.W. Freckman. 1997. The importance of marine biodiversity in ecosystem processes. *Ambio* 26:578–583.

Staddon, P.L., K. Thompson, I. Jakobsen, J.P. Grime, A.P. Askew, and A.H. Fitter. 2003. Mycorrhizal fungal abundance is affected by long-term climatic manipulations in the field. *Global Change Biology* 9:186–194.

Stone, M.K., and J.B. Wallace. 1998. Long-term recovery of a mountain stream from clear-cut logging: The effects of forest succession on benthic invertebrate community structure. *Freshwater Biology* 39:151–169.

Temple, S.A., and D.J. Flaspohler. 1998. The edge of the cut: Implications for wildlife populations. *Journal of Forestry* 96:22–26.

Termorshuizen, A., and A. Schaffers. 1991. The decline of carpophores of ectomycorrhizal fungi in stands of *Pinus sylvestris* L. in the Netherlands: Possible causes. *Nova Hedwigia* 53:267–289.

Thrush, S.F., 2004. Personal communication.

Thrush, S.F., J.E. Hewitt, A. Norkko, V.J. Cummings, and G.A. Funnell. 2003.

Catastrophic sedimentation on estuarine sandflats: Recovery of macrobenthic communities is influenced by a variety of environmental factors. *Ecological Applications* 13:1433–1455.

Thrush, S.F., C.J. Lundquist, and J.E. Hewitt. In press. Spatial and temporal scales of disturbance to the seafloor: A generalized framework for active habitat management. In: *Benthic Habitats and the Effects of Fishing*, edited by P.W. Barnes and J.P. Thomas. Bethesda, Maryland, American Fisheries, Symposium Series.

Topoliantz, S., J.F. Ponge, and P. Viaux. 2000. Earthworm and enchytraeid activity under different arable farming systems, as exemplified by biogenic structures. *Plant and Soil* 225:39–51.

Vallett, H.M., C.L. Crenshaw, and P.F. Wagner. 2002. Stream nutrient uptake, forest succession and biogeochemical theory. *Ecology* 83:2888–2901.

Vannote, R.L., G.W. Minshall, K.W. Cummins, J.R. Sedell, and C.E. Cushing. 1980. The river continuum concept. *Canadian Journal of Fisheries and Aquatic Sciences* 37:130–137.

Wall, D.H., M.A. Palmer, and P.V.R. Snelgrove. 2001a. Biodiversity in critical transition zones between terrestrial, freshwater, and marine soils and sediments: processes, linkages, and management implications. *Ecosystems* 4:418–420.

Wall, D.H., P.V.R. Snelgrove, and A.P. Covich. 2001b. Conservation priorities for soil and sediment invertebrates. In: *Conservation Biology*, edited by M.E. Soulé and G.H. Orians. Washington, DC, Island Press.

Wallace, J.B., and R.W. Merritt. 1980. Filter-feeding ecology of aquatic insects. *Annual Review of Entomology* 25:103–132.

Wall Freckman, D., T.H. Blackburn, L. Brussaard, P. Hutchings, M.A. Palmer, and P.V.R. Snelgrove. 1997. Linking biodiversity and ecosystem functioning of soils and sediments. *Ambio* 26:556–562.

Wallwork, J.A. 1970. *Ecology of Soil Animals*. London, McGraw-Hill.

Waters, T.F. 1995. *Sediment in Streams: Sources, Biological Effects, and Control*. Bethesda, Maryland, American Fisheries Society.

Webster, J.W., S.W. Golladay, E.F. Benfield, D.J. D'Angelo, and G.T. Peters. 1990. Effects of forest disturbance on particulate organic matter budgets of small streams. *Journal of the North American Benthological Society* 9:120–140.

Wheatcroft. R.A. 2000. Oceanic flood sedimentation: A new perspective. *Continental Shelf Research* 20:2059–2066.

Yanai, R.D., S.V. Stehman, M.A. Arthur, C.E. Prescott, A.J. Friedland, T.G. Siccama, and D. Binkley. 2003. Detecting change in forest floor carbon. *Soil Science Society of America Journal* 67:1583–1593.

10

Understanding the Functions of Biodiversity in Soils and Sediments Will Enhance Global Ecosystem Sustainability and Societal Well-Being

Diana H. Wall, Richard D. Bardgett, Alan P. Covich, and Paul V.R. Snelgrove

Many species of plants and animals live in soils and sediments, and they play crucial roles in providing ecosystem services for human well-being. A comprehensive synthesis of existing information on which habitats, taxa, and ecological functions in soil and freshwater and marine sediments are most essential is urgently needed if we are to maintain or restore their low-cost natural ecosystem services. This is of increasing importance when we recognize that more than 90 percent of the energy that flows through an ecosystem eventually passes through the food webs in the below-surface system. Therefore, the consequences of the loss of species and their functional roles may have far-reaching effects. Relatively little is known, however, about how species loss in soils and sediments will have direct and indirect effects on these functions and associated ecosystem services, or how this will feed back to below-surface systems (all chapters, this volume; Wall et al. 2001b).

This book addresses this vast underworld ecosystem and considers the consequences of global changes on the capacity of a rich diversity of organisms to provide ecosystem services. It is the first rigorous synthesis of the ability of the biodiversity both within and across soils and sediments to provide ecosystem services. To bring existing information together, international scientists specializing in the biodiversity and ecosystem functioning of one of the three domains examined the biota, habitats, and ecosystem functions provided by the biota. They further analyzed biotic interactions, abiotic factors, and the ecosystem services provided by the biota. This baseline was then incorporated into a preliminary appraisal of how the biota and services in each domain will be affected at var-

Table 10.1. Threats to soil biodiversity and to their ability to provide critical ecosystem services.

Category	Terrestrial	Freshwater	Marine
Exotic species	Imported soils	Ballast water	Ballast water
	Imported plants	Aquaculture	Aquaculture
Pollution	Mining	Agriculture	Sewage effluents
	Agriculture	Industrial waste	Oil spills
	Industrial waste	Aquaculture	Industrial waste
Harvesting	Agriculture	Water extraction	Fisheries
	Logging	Fisheries	Sand
			Ocean bottom trawling
Habitat destruction	Clearcutting	Water diversion	Coastal reclamation
	Agriculture	Dam building	Dredging
	Urbanization	Channelization	Fishing
Global climate change	Altered vegetation	Severe drought	Circulation changes
	Multiple climatic events of drought, rain	Severe flooding	Species compression
		Bank erosion	

ious spatial scales by global changes (e.g., land-use change, invasive species, atmospheric change) (see Table 10.1). The ecologists offer an in-depth appraisal of these issues by discussing specific case studies. They provide useful analytical frameworks that will enable readers to systematically consider the vulnerabilities of the organisms to global changes and the subsequent effects (positive or negative, direct or indirect) on other biota, the habitat, biogeochemical cycling, and ecosystem services. This approach provides powerful tools in the form of frameworks or templates that will allow scientists, land managers, and others to consider how understanding the below-surface system can provide management options for longer-term provision of ecosystem services.

This book fills a crucial gap in scientific knowledge by amplifying information on the critical roles of soil and sediment biota in the operation of the Earth's system. The observation that overexploiting one service can diminish another service emphasizes that trade-offs among alternative ecosystem management goals within and among domains can be important and should be considered in any analysis of complex ecosystem relationships. Linear approaches to management are often too focused on a single ecosystem service. For example, increasing crop or fish production by increased use of nitrogen-

and phosphorus-rich fertilizers often has detrimental downstream effects on water quality and recreational uses of lake and river ecosystems (Covich et al., Chapter 3; Giller et al., Chapter 6). Given these detailed and forward-thinking analyses about the sustainability of soils and sediments, what are some of the important findings?

A wide diversity of habitats exists within each domain. Soils and sediments are heterogeneous within small (centimeters to meter) and large (meter to kilometers) scales, with mosaics of physical and chemical habitats derived over geologic history and modified by climate, weathering (e.g., water, erosion), and above- and below-surface biodiversity. In other words, habitats within each of the soils, freshwater sediment, and marine sediment domains are often not physically or chemically alike. Also, the biological species that have evolved in these heterogeneous habitats have a range of life-history characteristics that determine their ability to withstand and recover from different global environmental changes.

The below-surface biota are intimately connected to other ecosystems. If we are to appreciate their role in providing ecosystem services, it is imperative that we integrate interdisciplinary research to examine these connections to adjacent ecosystems. Soils and sediment domains connect physically, chemically, and biologically, in three major ways: (1) below-surface, (2) above- and below-surface, and (3) laterally, above-surface (Table 10.2; Chapter 1, Figure 1.1). The biota in each domain, including soils, are fundamental to the regulation of groundwater quality and quantity and to the transfer of materials to adjacent domains. The food webs in the habitats of each domain become linked in a below-surface network that rapidly transforms and transfers materials, particles, and organisms across spatial scales that cover centimeters to kilometers. The feedback linkages between the below-surface components that control these material fluxes are understudied relative to above- and below-surface connections (Wardle et al. 2004). What is evident is that, for soils and sediments, the role of the biota in providing ecosystem services declines as human use intensifies. For example, the rapid increase in the rate of soil biotic habitats (soil types) (Amundson et al. 2003) and land area lost to urbanization and agricultural expansion is a major global-change driver affecting soil biodiversity and its provision of ecosystem services (van der Putten et al., Chapter 2; Wardle et al., Chapter 5; Wall et al. 2001a).

Another key finding is that there are similarities in the ecosystem processes (e.g., nutrient and energy pathways) (Table 10.2) and services, and in the functional groups (Chapter 1, Table 1.3), but the biodiversity of the organisms differs greatly across domains. Ecosystem processes such as decomposition, primary production, and hydrological cycling are major components of the below-surface system of sediments and soils, and bioturbators, or ecosystem engineers, appear to be critical taxa in many ecosystems for stabilizing both soil and sediments. Our knowledge, however, about the distribution, diversity, and role of larger organisms in driving ecosystem processes is better understood than it is for the smaller fauna.

Table 10.2. Below-surface connectivity and examples of similar ecosystem processes.

The connectivity of each domain (soil, freshwater sediments, or marine sediments) determines physical and biological states of adjacent terrestrial and aquatic ecosystems. The biota living below the surface regulate the movement and fate of materials and are integrally connected to physical and chemical environments and to other biota.

Interconnectivity of Below-surface Habitats	Similarities in Ecosystem Processes
Below surface	
Soils, freshwater sediments, and marine sediments are linked by feedbacks in various cycles	Primary productivity, leaching of nutrients, transfer of water, nutrients, particles and organisms; specialized biota in interface habitats
Each soil and sediment domain is linked to groundwater	Recapture/loss of nutrients, particles, and organisms from/to groundwater
Above and below surface	
Each soil and sediment component is linked to dynamic interfaces and to the atmosphere	Nutrient cycling through decomposition and primary productivity, secondary production; atmosphere connections; biodiversity and food web dependence
Lateral: above surfaces	
Ocean ↔ Freshwater Freshwater ↔ Land Land ↔ Ocean	Erosion and deposition of inorganic and organic particles, hydrologic cycling, energy transfers, nutrient cycling; transfer of organisms, and particles

Where Do We Go From Here?

There is an abounding biodiversity in soils and sediments, although we still do not know many of the species present or whether there are hot spots of biodiversity in different domains. This limitation is partly because much of our global knowledge of biodiversity and ecosystem functioning in soils, freshwater sediments, and ocean sediments is gained from studies primarily at small scales in northern temperate ecosystems. We have little evidence that biodiversity at the species level is directly related to ecosystem functioning, particularly for marine and freshwater sediment organisms where there have been fewer experiments than there have been for soils (Covich et al., Chapter 3; Weslawski et al., Chapter 4; Giller et al., Chapter 6; Snelgrove et al., Chapter 7). However, assemblages of species and their interactions in food webs appear to be key components in regulating ecosystem processes. Less understood is the extent to which specific interactions—such as facilitative interactions, symbioses of microbes with different metabolic functions, parasites of invertebrates, or predator-prey relationships—are critical to the provision of ecosystem services in or among soils and sediments. The resistance and recovery of assemblages of soil organisms to a disturbance or stress, or to multiple stresses such as with global changes, is also not well known and can be very context dependent. Studies of ecosystem functioning and the provision of ecosystem services need to include these and many abiotic variables that affect biodiversity patterns and their linkages within and across below-surface domains to establish a more comprehensive understanding for sustainable management and conservation. Quantitative information gained at the multi-species level from a number of robust experiments at small and large spatial scales and longer temporal scales must be conducted regionally and globally for a greater predictive capability concerning threats to, and controls on, different ecosystems and their services.

Global changes affect soil and sediment organisms, directly and indirectly, through connections to other ecosystems (Table 10.2). There is sufficient evidence to indicate that diversity of species decreases with habitat disturbance (climatic or direct human intervention), leading to the dominance of communities by a few species. Therefore, measures of biodiversity other than species richness, such as evenness of species abundances, are needed as indicators of disturbances that impact soil and sediment biodiversity with a subsequent loss of services. Different, but relevant, indicators of soil and sediment biodiversity are critical if we are to expand management options, which presently are based on relatively few species in microcosm and small-scale field experiments, to more diverse assemblages and to larger spatial and longer temporal scales. Because organisms and their regulation of ecosystem processes occur across a range of scales within and among domains, the impact of global changes will be manifested on ecosystem services across multiple scales (Lavelle et al., Chapter 8; Ineson et al., Chapter 9). A priority for future research is incorporating effects of multiple stressors into experiments regionally.

The information presented here provides sufficient evidence that soils and sediments and their biota must no longer be considered as isolated systems that are less urgently in need of study. It is imperative for ecosystem sustainability at local, regional, and global levels that all of society considers the vertebrate, invertebrate, and microbial biodiversity of soils and sediments as having a crucial role in the provision of ecosystem services. It is apparent, based on the brief summary above, that in order to determine whether valued below-surface ecosystem services will be lost under global changes, we must determine through a quantitative, systematic approach how critical species respond to disturbances, and identify which biota and processes are most sensitive.

Literature Cited

Wall, D.H., G.A. Adams, and A.N. Parsons. 2001a. Soil biodiversity. In: *Global Biodiversity in a Changing Environment: Scenarios for the 21st Century*, edited by F.S. Chapin, III, O.E. Sala, and E. Huber-Sannwald, pp. 47–82. New York, Springer-Verlag.

Wall, D.H, P.V.R. Snelgrove, and A.P. Covich. 2001b. Conservation priorities for soil and sediment invertebrates. In: *Conservation Biology*, edited by M.E. Soulé and G.H. Orians, pp. 99–123. Washington, DC, Island Press.

Wardle, D.A., R.D. Bardgett, J.N. Klironomos, H. Setälä, W.H. van der Putten, and D.H. Wall. 2004. Ecological linkages between aboveground and belowground biota. *Science* 304:1629–1633.

Contributors

Chair, SCOPE Committee on Soil and Sediment Biodiversity and Ecosystem Functioning (SSBEF)

DIANA H. WALL is a professor and Director of the Natural Resource Ecology Laboratory, an international ecosystem research center at Colorado State University, and is a former president of the Ecological Society of America, the American Institute of Biological Sciences, and the Society of Nematologists. Her research addresses the importance of soil biodiversity for ecosystems and the consequences of human activities on soil sustainability. She currently studies how soil biodiversity contributes to ecosystems from the limited diversity of the Antarctic Dry Valleys to the ecosystems of higher biodiversity.

Co-Chairs

RICHARD D. BARDGETT served as chair of the Soil Domain for the SSBEF meeting in Estes Park, Colorado, and is a professor of ecology in the Department of Biological Sciences at Lancaster University in the UK. His research aims to determine how changes in environmental conditions alter the size, diversity, and activity of soil biological communities, and, in turn, how changes in soil biological properties influence the character and functioning of natural and managed ecosystems.

ALAN P. COVICH served as the chair of the Freshwater Domain for the SSBEF meeting in Estes Park, Colorado, and is the director of the Institute of Ecology at the University of Georgia. He is a former president of the American Institute of Biological Sciences and the North American Benthological Society. His research focuses on aquatic food webs in temperate and tropical streams, and the effects of disturbances on predator-prey dynamics, with research sites in Mexico, Central America, and the Caribbean.

PAUL V.R. SNELGROVE served as the chair of the Marine Domain for the SSBEF meeting in Estes Park, Colorado, and is an associate professor and Canada Research Chair in Boreal and Cold Ocean Systems at Memorial University of Newfoundland. He studies the role of transport of larval fish and invertebrates and how these contribute to patterns in marine benthic communities, as well as in the factors that help regulate species diversity in marine environments.

Meeting Participants

JONATHAN M. ANDERSON is a professor of ecology in the Department of Biological Sciences at the University of Exeter, UK. He was a co-founder of the Tropical Soil Biology and Fertility Program (now TSBF-CIAT) and has served on the Sciences of Soils Advisory Board. His research interests include soil biodiversity and ecosystem functioning, management of soil processes, and agrobiodiversity in tropical farming systems.

MELANIE C. AUSTEN leads a research group on Biodiversity and Ecosystem Functioning at Plymouth Marine Laboratory, UK. Her research interests include links between biodiversity and ecosystem function, impact of fishing on marine ecosystem processes and goods and services, socioeconomic importance of marine biodiversity, field and experimental benthic ecology in coastal habitats, nematode taxonomy using traditional and molecular techniques, and benthic-pelagic coupling.

VALERIE BEHAN-PELLETIER is a research scientist for Agriculture and Agrifood Canada, and a visiting lecturer at the Acarology Summer Program, Ohio State University. Valerie is a recipient of the Japanese Society for the Promotion of Science Fellowship. She has served as executive secretary for the International Congress of Acarology, on the Scientific Committee of the Biological Survey of Canada, and as a member of the Editorial Board of the *Soil Mites of the World* series. Her research interests include oribatid mite systematics, biodiversity and ecology in tropical rainforest, grasslands, and canopy habitats.

DAVID E. BIGNELL is a professor of zoology in the School of Biological Sciences, Queen Mary, University of London, UK. David serves on the Steering Committee of the Global Litter Invertebrate Decomposition Experiment (GLIDE), and has published over 80 refereed papers and reviews on the biology of locusts, cockroaches, millipedes, and termites, with the main focus on gut structure, digestive physiology, intestinal microbiology, nutritional ecology, and field ecology. Currently, his research focuses on the measurement of soil biodiversity and the assessment of the role of soil organisms in ecosystem processes.

GEORGE G. BROWN is a researcher at Embrapa Soybean, in Londrina, Brazil. George has participated in and led several international projects and networks on soil macrofauna, with research in Latin America and Africa. His research interests include earthworm ecology and biodiversity, role of soil biota in sustainable agriculture, and use of biological indicators of soil quality.

VALERIE K. BROWN is professor of agro-ecology and director of the Centre for Agri-Environmental Research at the University of Reading, UK. She is a member of the UK

Natural Environment Research Council and the Scientific Advisory and Research Priority Groups to DEFRA (Department for Environment, Food and Rural Affairs). She currently holds positions in IGBP and GCTE and has served as vice president of the British Ecological and the Royal Entomological Societies. Her research interests include multitrophic interactions, above-belowground biotic interactions, insect-fungal interactions, land-use change, agro-ecology, and climate change.

LIJBERT BRUSSAARD is a professor in the Department of Soil Quality at Wageningen University, The Netherlands. He researches the role of soil fauna in decomposition processes and nutrient cycling; the relationships between below- and aboveground biodiversity and ecosystem functioning in agro(forestry)-ecosystems; and soil quality assessment using soil fauna.

KATHERINE C. EWEL is a research ecologist and senior scientist at the Institute of Pacific Islands Forestry, Pacific Southwest Research Station, USDA Forest Service, in Honolulu, Hawaii. She conducts research on the ability of mangrove forests and freshwater forested wetlands in the Pacific islands to provide goods and services to human populations. This includes characterizing forest structure and hydrologic relationships, evaluating the impacts of resource utilization, and understanding the socioeconomic context for resource management.

JAMES R. GAREY is an associate professor of biology at the University of South Florida. His research is directed toward understanding the evolutionary relationships of the major groups of invertebrates, with a focus on lesser-known and meiofaunal groups, using a combination of molecular and morphological characters. He applies molecular phylogenetic methods to the study of meiofaunal community structure using high throughput DNA analysis. Jim also examines the biogeochemistry of coastal karst cave systems stemming from his interest in cave diving.

PAUL S. GILLER is a professor in the Department of Zoology, Ecology and Plant Science and is currently in his second term as dean of science at the University College Cork, Ireland. Paul is an ecologist with expertise in community ecology and freshwater biology, particularly in the analysis of macroinvertebrate communities of freshwater habitats, freshwater-forestry interactions, fish diet and feeding strategies, the impact of instream and catchment disturbances on stream and river ecosystems, and the role of diversity on ecosystem function. He has also worked on terrestrial ecosystems, particularly dung beetle communities and semi-natural grasslands.

WILLEM GOEDKOOP is an associate professor at the Department of Environmental Assessment of the Swedish University of Agricultural Sciences. His main research interests are in invertebrate feeding biology, and the ecological linkages between pelagic and benthic communities and between aquatic and terrestrial environments. His most recent research focus includes the bioavailability of sediment contaminants in benthic food webs, in particular the interactions between contaminants and sediment microbes for contaminant bioconcentration and bioaccumulation.

ROBERT O. HALL, JR., is an assistant professor in the Department of Zoology and Physiology at the University of Wyoming. His research interests include the interaction between animal assemblages and ecosystem function in streams, energy and nutrient flow in food webs, invasive species impact to food webs, stable isotopes as food web tracers, nitrogen cycling in streams, and bacterivory by aquatic invertebrates.

STEPHEN J. HAWKINS is professor and director of the Marine Biological Association, Plymouth UK, and professor of environmental biology at the University of Southampton, School of Biological Sciences. Stephen conducts research on rocky shore community ecology; behavioral ecology of intertidal grazers; restoration of degraded coastal ecosystems; recovery of polluted shores and estuaries; long-term change in relation to climate using rocky-shore indicators; shellfisheries, impacts of scallop dredging on benthos; ecology and design of sea defenses; and taxonomy of intertidal gastropods —particularly Patellidae.

H. WILLIAM HUNT is a senior scientist and professor of rangeland ecosystem science in the Natural Resource Ecology Laboratory at Colorado State University, and an Long Term Ecological Research Faculty Associate. His areas of research include nutrient cycling, soil food webs, simulation models, nonlinear dynamics, systems ecology, and the effects of climate change on ecosystem function.

THOMAS M. ILIFFE is professor of marine biology at Texas A&M University at Galveston. Tom has discovered more than 200 new species and many new orders and genera of marine and freshwater cavernicolous invertebrates. His research interests include biodiversity, biogeography, evolution and ecology of animals inhabiting anchialine caves; cave conservation and environmental protection; and cave and research diving.

PHILIP INESON is professor of global change ecology in the Department of Biology at the University of York, UK, and is also a research group leader within the Stockholm Environment Institute at York. Phil is the 1989 winner of the Paulo Buchner International Award for Terrestrial Ecology, and has served on numerous scientific committees, including evaluations for the UK Natural Environment Research Council, European Union and the Academy of Finland. His research examines the impacts of global change and air pollution on ecosystems, with particular emphasis on soil organisms. The work ranges from the assessment of climate change impacts on natural ecosystems to characterizing the soil organisms important in trace gas release and uptake.

T. HEFIN JONES is a scientist with the Biodiversity and Ecological Processes Research Group in the Cardiff School of Biosciences, UK. Hefin is an honorary research fellow at the UK NERC Centre for Population Biology, Imperial College; an editor of *Global Change Biology;* and a member of the Editorial Board of Agricultural and Forest Entomology, Bulletin of Entomological Research, Biologist and Functional Ecology. He serves on the External Advisory Board for the Phytotron National Facility at Duke University, USA, and is a council member of both the British Ecological Society and the

Royal Entomological Society. His research interests encompass population and community ecology, climate change, and biodiversity, centering on three main areas: host-parasitoid interactions, soil biodiversity, and climate change.

RONALD T. KNEIB is a senior research scientist at the University of Georgia Marine Institute and an adjunct associate professor of marine sciences at the University of Georgia. He is certified by the Ecological Society of America as a senior ecologist, served for over 10 years on the editorial board of Marine Ecology Progress Series, and is a former secretary-treasurer for the Southeastern Estuarine Research Society. Ron's research focuses on population and community dynamics of benthic and epibenthic fishes and invertebrates in tidal wetlands, the effects of spatial patterns in marsh landscapes on ecological processes and functions measured across scales from genomes to ecosystems.

PATRICK LAVELLE is a professor of ecology at Université de Paris VI and director of the BIOSOL Unit at Institut de Recherche sur le Développement. He is a member of the France National Academy of Sciences. His research interests include general soil ecology with a special emphasis set on earthworm ecology and their management as part of sustainable practices in tropical environments, relationships between above- and belowground diversity, consequences for soil function and plant growth, and use of soil macrofauna communities as bioindicators of soil quality.

LISA A. LEVIN is a professor and research scientist for the Marine Life Research Group at the Scripps Institution of Oceanography. She studies the ecology of soft-sediment assemblages in wetlands, estuaries, continental margins, and the deep sea; animal-sediment interactions; consequences of species invasion; ecology of methane seeps; oxygen minimum zones; and salt marsh restoration.

DAVID M. MERRITT is a riparian plant ecologist at the National Stream Systems Technology Center, USFS Rocky Mountain Research Station, and a visiting scientist at the Natural Resource Ecology Laboratory at Colorado State University. David is an alumni of the Nature Conservancy David H. Smith Conservation Fellowship Program, former president of the Rocky Mountain Chapter of the Society of Wetland Scientists, and a member of the Landscape Ecology Group at Umeå University, Sweden. His research focuses on the factors influencing plant species diversity and plant invasions in riparian ecosystems with an emphasis on the role of river damming and water development on such processes.

ELDOR A. PAUL is a research scientist at the Natural Resource Ecology Laboratory, Colorado State University, and professor emeritus at Michigan State University. His textbook *Soil Microbiology and Biochemistry*, coauthored with F.E. Clark, is widely read in ecosystem science, agronomy, and soil science and has been translated into three languages. Eldor's research interests have centered on the ecology of soil biota, the role of nutrients such as N in plant growth, and the dynamics of C and N in sustainable agriculture and global change.

MARK ST. JOHN is a Ph.D. candidate in the Graduate Degree Program in Ecology at the Natural Resource Ecology Laboratory, Colorado State University. He studies interactions between plants and soil mites and their influences on decomposition at the Konza Prairie Long Term Ecological Research Site, Kansas, USA.

WIM H. VAN DER PUTTEN is department head at the Centre for Terrestrial Ecology of the Netherlands Institute of Ecology and Professor at Wageningen University. He studies soil multitrophic interactions in relation to spatio-temporal processes in natural vegetation; linking to aboveground multitrophic interactions; plant invasiveness in relation to above- and belowground trophic interactions; natural succession; ecology, regulation, and host specificity of plant parasitic nematodes in natural ecosystems; plant-microorganism interactions; and functional biodiversity.

DAVID A. WARDLE is a professor at the Swedish University of Agricultural Sciences at Umeå, Sweden, and an ecologist at Landcare Research in New Zealand. He is the author of the book *Communities and Ecosystems: Linking the Aboveground and Belowground Components*, recently published in the Princeton University Press Monographs in the Population Biology series. Most of his work is focused on rainforests in New Zealand and boreal forests in northern Sweden. His most recent projects involve the use of islands as model systems for understanding drivers of ecosystems.

JAN MARCIN WESLAWSKI is professor and head of the Department of Marine Ecology at the Institute of Oceanology, Polish Academy of Sciences in Sopot. He is a member of the Arctic Ocean Sciences Board, as well as a member of the Polish Scientific Committee on Oceanic Research. His research interests include ecology of crustaceans and the functioning of Arctic marine ecosystems.

ROBERT B. WHITLATCH is a professor in the Department of Marine Sciences and Ecology and Evolutionary Biology at the University of Connecticut. His research focuses on marine benthic population and community ecology, using laboratory and field experimentation, in combination with modeling, to address how abiotic and biotic processes influence the distribution and composition of populations and communities.

TODD WOJTOWICZ is a Ph.D. candidate in the Graduate Degree Program in Ecology at the Natural Resource Ecology Laboratory, Colorado State University. He is currently studying the influence of plant species identity on soil communities in a semi-arid grassland system at the shortgrass-steppe LTER.

HOLLEY ZADEH is a research associate in the Natural Resource Ecology Laboratory, Colorado State University. She has a post-graduate degree in Restoration Ecology from the Rangeland Ecosystem Sciences Department at CSU, and currently serves as the project director for the Global Litter Invertebrate Decomposition Experiment (GLIDE) and as the SCOPE SSBEF publication coordinator in Diana Wall's Soil and Sediment Biodiversity and Ecosystem Functioning group at NREL.

SCOPE Series List

SCOPE 47: *Long-Term Ecological Research: An International Perspective*, 1991, 312 pp

SCOPE 48: *Sulphur Cycling on the Continents: Wetlands, Terrestrial Ecosystems and Associated Water Bodies*, 1992, 345 pp

SCOPE 49: *Methods to Assess Adverse Effects of Pesticides on Non-target Organisms, SGOMSEC 7*, 1992, 264 pp

SCOPE 50: *Radioecology After Chernobyl*, 1993, 367 pp

SCOPE 51: *Biogeochemistry of Small Catchments: A Tool for Environmental Research*, 1993, 432 pp

SCOPE 52: *Methods to Assess DNA Damage and Repair: Interspecies Comparisons, SGOMSEC 8*, 1994, 257 pp

SCOPE 53: *Methods to Assess the Effects of Chemicals on Ecosystems, SGOMSEC 10*, 1995, 440 pp

SCOPE 54: *Phosphorus in the Global Environment: Transfers, Cycles and Management*, 1995, 480 pp

SCOPE 55: *Functional Roles of Biodiversity: A Global Perspective*, 1996, 496 pp

SCOPE 56: *Global Change: Effects on Coniferous Forests and Grasslands*, 1996, 480 pp

SCOPE 57: *Particle Flux in the Ocean*, 1996, 396 pp

SCOPE 58: *Sustainability Indicators: A Report on the Project on Indicators of Sustainable Development*, 1997, 440 pp

SCOPE 59: *Nuclear Test Explosions: Environmental and Human Impacts*, 1999, 304 pp

SCOPE 60: *Resilience and the Behavior of Large-Scale Systems*, 2002, 287 pp

SCOPE 61: *Interactions of the Major Biogeochemical Cycles: Global Change and Human Impacts*, 2003, 358 pp

SCOPE 62: *The Global Carbon Cycle: Integrating Humans, Climate, and the Natural World*, 2004, 584 pp

SCOPE 63: *Invasive Alien Species: A New Synthesis*, 2004, 352 pp

SCOPE 64: *Sustaining Biodiversity and Ecosystem Services in Soils and Sediments*, 2004, 304 pp

SCOPE 65: *Agriculture and the Nitrogen Cycle*, edited by Arvin R. Mosier, J. Keith Syers, and John R. Freney, 320 pp

SCOPE Soil and Sediment Biodiversity and Ecosystem Functioning (SSBEF) Committee Publications

Adams, G.A., and D.H. Wall. 2000. Biodiversity above and below the surface of soils and sediments: Linkages and implications for global change. *BioScience* 50:1043–1048.

Adams, G.A., D.H. Wall, and A.P. Covich. 1999. Linkages between below-surface and above-surface biodiversity. *Bulletin of the Ecological Society of America* 80:200–204.

Austen, M.C., P.J.D. Lambshead, P.A. Hutchings, G. Boucher, P.V.R. Snelgrove, C. Heip, G. King, I. Koike, and C. Smith. 2002. Biodiversity links above and below the marine sediment-water interface that may influence community stability. *Biodiversity and Conservation* 11:113–136.

Bardgett, R.D., J.M. Anderson, V. Behan-Pelletier, L. Brussaard, D.C. Coleman, C. Ettema, A. Moldenke, J.P. Schimel, and D.H. Wall. 2001. The influence of soil biodiversity on hydrological pathways and the transfer of materials between terrestrial and aquatic ecosystems. *Ecosystems* 4:421–429.

Behan-Pelletier, V., and G. Newton. 1999. Computers in biology: Linking soil biodiversity and ecosystem function—The taxonomic dilemma. *BioScience* 49:149–153.

Brussaard, L., V.M. Behan-Pelletier, D.E. Bignell, V.K. Brown, W. Didden, P. Folgarait, C. Fragoso, D. Wall Freckman, V.V.S.R. Gupta, T. Hattori, D.L. Hawksworth, C. Klopatek, P. Lavelle, D.W. Malloch, J. Rusek, B. Soderstrom, J.M. Tiedje, and R.A. Virginia. 1997. Biodiversity and ecosystem functioning in soil. *Ambio* 26:563–570.

Covich, A.P., M.A. Palmer, and T.A. Crowl. 1999. The role of benthic invertebrate species in freshwater ecosystems: Zoobenthic species influence energy flows and nutrient cycling. *BioScience* 49:119–127.

Ewel, K.C., C. Cressa, R.T. Kneib, P.S. Lake, L.A. Levin, M. Palmer, P. Snelgrove, and D.H. Wall. 2001. Managing critical transition zones. *Ecosystems* 4:452–460.

Finlay, B.J., and G.F. Esteban. 1998. Freshwater protozoa: Biodiversity and ecological function. *Biodiversity and Conservation* 7:1163–1186.

Folgarait, P.J. 1998. Ant biodiversity and its relationship to ecosystem functioning: A review. *Biodiversity and Conservation* 7:1221–1244.

Groffman, P.M., and P.J. Bohlen. 1999. Soil and sediment biodiversity: Cross-system comparisons and large-scale effects. *BioScience* 49:139–148.

Hooper, D.U., D.E. Bignell, V.K. Brown, L. Brussaard, J.M. Dangerfield, D.H. Wall, D.A. Wardle, D.C. Coleman, K.E. Giller, P. Lavelle, W.H. van der Putten, P.C. de

Ruiter, J. Rusek, W.L. Silver, J.M. Tiedje, and V. Wolters. 2000. Interactions between aboveground and belowground biodiversity in terrestrial ecosystems: Patterns, mechanisms, and feedbacks. *BioScience* 50:1049–1061.

Hutchings, P. 1998. Biodiversity and functioning of polychaetes in benthic sediments. *Biodiversity and Conservation* 7:1133–1145.

Hyde, K.D., E.B. Gareth Jones, E. Leaño, S.B. Pointing, A.D. Poonyth, and L.L.P. Vrijmoed. 1998. Role of fungi in marine ecosystems. *Biodiversity and Conservation* 7:1147–1161.

Ingram J., and D. Wall Freckman. 1998. Soil biota and global change: Preface. *Global Change Biology* 4:699–701.

Lake, P.S., M.A. Palmer, P. Biro, J. Cole, A.P. Covich, C. Dahm, J. Gibert, W. Goedkoop, K. Martens, and J. Verhoeven. 2000. Global change and the biodiversity of freshwater ecosystems: Impacts on linkages between above-sediment and sediment biota. *BioScience* 50:1099–1107.

Lavelle, P., D. Bignell, M. Lepage, V. Wolters, P. Roger, P. Ineson, O.W. Heal, and S. Dhillion. 1997. Soil function in a changing world: The role of invertebrate ecosystem engineers. *European Journal of Soil Biology* 33:159–193.

Levin, L.A., D.F. Bosch, A. Covich, C. Dahm, C. Erseus, K.C. Ewel, R.T. Kneib, A. Moldenke, M.A. Palmer, P. Snelgrove, D. Strayer, and J.M. Weslawski. 2001. The function of marine critical transition zones and the importance of sediment biodiversity. *Ecosystems* 4:430–451.

Palmer, M., A.P. Covich, B.J. Finlay, J. Gibert, K.D. Hyde, R.K. Johnson, T. Kairesalo, S. Lake, C.R. Lovell, R.J. Naiman, C. Ricci, F. Sabater, and D. Strayer. 1997. Biodiversity and ecosystem processes in freshwater sediments. *Ambio* 26:571–577.

Palmer, M.A., A.P. Covich, S. Lake, P. Biro, J.J. Brooks, J. Cole, C. Dahm, J. Gibert, W. Goedkoop, K. Martens, J. Verhoeven, and W.J. van de Bund. 2000. Linkages between aquatic sediment biota and life above sediments as potential drivers of biodiversity and ecological processes. *BioScience* 50:1062–1075.

Rusek, J. 1998. Biodiversity of Collembola and their functional role in the ecosystem. *Biodiversity and Conservation* 7:1207–1219.

Schimel, J.P., and J. Gulledge. 1998. Microbial community structure and global trace gases. *Global Change Biology* 4:745–758.

Smith, P., O. Andren, L. Brussaard, M. Dangerfield, K. Ekschmitt, P. Lavelle, and K. Tate. 1998. Soil biota and global change at the ecosystem level: Describing soil biota in mathematical models. *Global Change Biology* 4:773–784.

Smith, C.R., M.C. Austen, G. Boucher, C. Heip, P.A. Hutchings, G.M. King, I. Koike, P.J.D. Lambshead, and P. Snelgrove. 2000. Global change and biodiversity linkages across the sediment-water interface. *BioScience* 50:1108–1120.

Snelgrove, P.V.R. 1998. The biodiversity of macrofaunal organisms in marine sediments. *Biodiversity and Conservation* 7:1123–1132.

Snelgrove, P.V.R. 1999. Getting to the bottom of marine biodiversity: Sedimentary habitats—ocean bottoms are the most widespread habitat on Earth and support high biodiversity and key ecosystem services. *BioScience* 49:129–138.

Snelgrove, P.V.R., M.C. Austen, G. Boucher, C. Heip, P.A. Hutchings, G.M. King, I. Koike, P.J.D.Lambshead, and C.R. Smith. 2000. Linking biodiversity above and below the marine sediment-water interface. *BioScience* 50:1076–1088.

Snelgrove, P, T.H. Blackburn, P.A. Hutchings, D.M. Alongi, J.F. Grassle, H. Hummel, G.

King, I. Koike, P.J.D. Lambshead, N.B. Ramsing, and V. Solis-Weiss. 1997. The importance of marine sediment biodiversity in ecosystem processes. *Ambio* 26:578–583.

Swift, M.J., O. Andren, L. Brussaard, M. Briones, M.-M. Couteaux, K. Ekschmitt, A. Kjoller, P. Loiseau, and P. Smith. 1998. Global change, soil biodiversity and nitrogen cycling in terrestrial ecosystems: Three case studies. *Global Change Biology* 4:729–743.

van Noordwijk, M., P. Martikainen, P. Bottner, E. Cuevas, C. Rouland, and S.S. Dhillion. 1998. Global change and root function. *Global Change Biology* 4:759–772.

Wall, D.H. 1999. Biodiversity and ecosystem functioning. *BioScience* 49:107–108.

Wall, D.H., G.A. Adams, and A.N. Parsons. 2001. Soil Biodiversity. In: *Global Biodiversity in a Changing Environment: Scenarios for the 21st Century*, edited by F.S. Chapin III and O.E. Sala, pp. 47–82. New York, Springer-Verlag.

Wall, D.H., L. Brussaard, P.A. Hutchings, M.A. Palmer, and P.V.R. Snelgrove. 1998. Soil and sediment biodiversity and ecosystem functioning. *Nature and Resources* 34:41–51.

Wall, D.H., and J.C. Moore. 1999. Interactions underground: Soil biodiversity, mutualism, and ecosystem processes. *BioScience* 49:109–117.

Wall, D.H., M.A. Palmer, and P.V.R. Snelgrove. 2001. Biodiversity in critical transition zones between terrestrial, freshwater, and marine soils and sediments: Processes, linkages, and management implications. *Ecosystems* 4:418–420.

Wall, D.H., P.V.R. Snelgrove, and A. Covich. 2001. Conservation priorities for soil and sediment invertebrates. In: *Conservation Biology: Research Priorities for the Next Decade*, edited by M.E. Soulé and G.H. Orians, pp. 99–123. Washington, DC, Island Press.

Wall Freckman, D., T.H. Blackburn, L. Brussaard, P. Hutchings, M.A. Palmer, and P.V.R. Snelgrove. 1997. Linking biodiversity and ecosystem functioning of soils and sediments. *Ambio* 26:556–562.

Wardle, D.A. 1999. Biodiversity, ecosystems and interactions that transcend the interface. *Trends in Ecology and Evolution* 14:125–127.

Wardle, D.A., H.A. Verhoef, and M. Clarholm. 1998. Trophic relationships in the soil microfood-web: Predicting the responses to a changing global environment. *Global Change Biology* 4:713–727.

Wolters, V., W.L. Silver, D.E. Bignell, D.C. Coleman, P. Lavelle, W.H. van der Putten, P. de Ruiter, J. Rusek, D.H. Wall, D.A. Wardle, L. Brussaard, J.M. Dangerfield, V.K. Brown, K.E. Giller, D.U. Hooper, O. Sala, J. Tiedje, and J.A. van Veen. 2000. Effects of global changes on above- and belowground biodiversity in terrestrial ecosystems: Implications for ecosystem functioning. *BioScience* 50:1089–1098.

Wong, M.K.M., T.-K. Goh, I.J. Hodgkiss, K.D. Hyde, V.M. Ranghoo, C.K.M. Tsui, W.-H. Ho, W.S.W. Wong, and T.-K. Yuen. 1998. Role of fungi in freshwater ecosystems. *Biodiversity and Conservation* 7:1187–1206.

Young, I.M., E. Blanchart, C. Chenu, M. Dangerfield, C. Fragoso, M. Grimaldi, J. Ingram, and L.J. Monrozier. 1998. The interaction of soil biota and soil structure under global change. *Global Change Biology* 4:703–712.

SCOPE Executive Committee 2001–2004

President
Dr. Jerry M. Melillo (USA)

1st Vice-President
Prof. Rusong Wang (China-CAST)

2nd Vice-President
Prof. Bernard Goldstein (USA)

Treasurer
Prof. Ian Douglas (UK)

Secretary-General
Prof. Osvaldo Sala (Argentina-IGBP)

Members
Prof. Himansu Baijnath (South Africa-IUBS)
Prof. Manuwadi Hungspreugs (Thailand)
Prof. Venugopalan Ittekkot (Germany)
Prof. Holm Tiessen (Canada)
Prof. Reynaldo Victoria (Brazil)

Index